VOICES OF THE EARTH

LIFE LESSONS FROM A BLUE PLANET

BY ADAM CHAPUIS

Proudly Published in the USA by

ISBN: 1519278233

Cover Design by:
Capri Brock

ACKNOWLEDGEMENTS

This book began as a collection of thoughts along a life path of embracing science and spirituality. The hard science was gathered in institutions of learning, in poking around the Earth while practicing geology as a craft and in sharing thoughts around many campfires. The spirituality came more slowly, a series of connections and "ah ha" moments along many spiritual paths, particularly the Science of Mind philosophy.

But it never would have gone any further than guiding people on a "Spiritual Walk on Dinosaur Ridge" without the inspiration, unwavering support and hard questions of Robin Rose who just never doubted for a moment that it could be done.

Ideas need support and nurturing to round and smooth them out and there were many. But I would especially thank William Robert Case and Simon Shadowlight. We often met on Fridays to listen to each other's thoughts and to remember our commitments to our writing. I would also acknowledge the entire group of wonderful men I worked and laughed with in the Men's Ministry of the Mile High Church of Denver. Supporting those activities provided an avenue of self exploration and it remains one of the most rewarding things I have ever been a part of.

A very special and heartfelt thanks goes out to Aeeshah and Kokomon Clottey, and Keya Kessler, who provided wisdom and inspiration as fellow travelers who care deeply about the Earth and the children who call it home.

And there was another constant companion who shared my mountain land with me, who sat as still as I did listening to the power of nature and breathing in all that is Earth. My beloved Samoyed, Manna, has since passed on but she laughed and loved as only a dog can do, and made the very most of her time on this planet.

*You can never write for the whole world,
you write in hopes the world will become whole.*

DEDICATION

This book is dedicated to all those souls out there dancing lightly with Earth – those who steal mysticism by moonlight, who take pleasure in the power of a thunderstorm and rejoice in its cleansing rain.

It is for those who love life in all of its expressions and sometimes shed a tear at the closeness of Spirit and its overpowering beauty. And it is to those who have wonderful things yet unfinished, things yet to be acknowledged and yet to be unfolded.

We carry the history of the Earth in our DNA and perhaps in our souls. When we see the Earth in this fashion we begin to understand our place in this drama.

TABLE OF
CONTENTS

INTRODUCTION

About ten minutes from where I live there is a most wondrous place. I know it well because I am fortunate enough to pass through its domain, feel its imposing presence on my way to work each day and then again returning home. My journey begins on the western edge of Denver, Colorado where I take a lightly traveled road westward out of the last rolling hills of the Great Plains. As I crest a low hill I look out across a valley and directly back in time to what is now called Dinosaur Ridge. Its rising up is so abrupt, and it forms such an obstacle to the mind that I fight the instant urge to slow the car down as if the road will soon end though I am yet a mile away. As I drive on I recognize that the Ridge is part of a north/south line of demarcation that marks the beginnings of the front range of the Rocky Mountains.

But there is more, for in my vision beyond and over the Ridge, and separated by a hidden valley, is the beautiful array of brilliantly red and steeply dipping sandstones that form Red Rocks Park. In this sandstone lies Red Rocks Amphitheater, where performers and spectators celebrate among rocks deposited in the Age of Reptiles. Some smile and think it is a fitting place for the sound of good old rock and roll that often fills a summer night. Slightly further west my attention is drawn to a pinnacled mountain edging its way above the rest, for the strong and stark rocks of the amphitheater are nestled on the broad eastern shoulders of Mount Morrison, a landmark rising to 8,000 feet above sea level and composed of 1.5 billion-year-old sediments. It reminds me of the story of Atlas for surely Mount Morrison holds up this part of the world. Yet the mountain is only a smaller brother in the family of Rocky Mountains that rises up as individuals, each progressively higher to the west.

There is another more recent story that is preserved in the sediments that lie along the Ridge in front of me. It is a story of dinosaurs and an ancient beachfront property; it's about change and timeless principles; it's about echoes from the past and lessons for the future. And it is this story that has captured my imagination.

One hundred million years ago this area was a beach much like parts of Florida are today. Shallow water and low-energy waves lapped along a subtle coastline marked by elaborate mangrove swamps. It was along this sandy coast that a variety of dinosaurs wandered under a subtropical sun, each exploring; seeking food and company, and each joined completely in the wonder and the struggle that is life. The ancient mangroves with their roots and stems; the old stream beds that fed into the sea, some with apparent scratches still visible from the claws of crocodiles; the ripple marks from wave action just off the ancient shore and the footprints of the dinosaurs themselves are all wonderfully preserved, as if alive and active only yesterday. Over time the climates changed, the ocean withdrew, and of course the dinosaurs became extinct. And this snapshot, this detailed picture in the rock with its messages like the Ten Commandments, was covered over in the relentless evolution of mountain building and erosion.

Eventually the Earth released its buried treasure of dinosaur footprints in its own subtle and persistent fashion. Long ago the sediments of the Ridge were deposited by an array of streams and then sculpted and reworked along the edges of a shallow sea. Although these sediments were originally flat, they have since been raised up in the tortured birth of the Rocky Mountains. Today they dip very sharply to the east and appear to almost stand on end, thin beds and thick, like a deck of cards. Each night during the winter, moisture from snow or rain that has collected in the small openings between the beds begins to freeze. As it becomes ice it expands with a subtle but very powerful force.

In the sunlight of the next day, it thaws again and that evening new moisture collects in the widened cracks. This continual freezing and thawing gradually pushed the rock apart, relentlessly weakening its bonds and loosening its grip until one day a large slab broke free and slid down the ridge, exposing the underlying footprints in the rock. So it stands today, available for our exploration and our communion some 36 billion sunsets after the soft sand was imprinted with the foot of the dinosaur.

Back in present time, I relax my focus. And in broadening my view, I notice that the Ridge rises up defiantly north and south as far as the eye can see as if to say to the adjacent gentle, soft-shouldered and dust-colored hills of the Great Plains that this is enough, this is your high-water mark, as far as you will go. From here on, the Earth will paint its pictures in sharp colors and powerful outlines. Much will change from this point west—the rock, the animals and the vegetation. The Earth seems to recognize this with a deep and thoughtful consciousness that envelopes all that stand and look.

As I near the Ridge, I turn at the last moment and follow a little road north along its base. Here the Ridge stands about 600 feet above me and at times seems almost to reach over me, the upper portion near its top exposed in stretches of bare cliff rock, but quickly covered in soft grass along the lower slopes. It looks like a giant wave rolling in from the west, caught at that precipitous moment when everyone knows the game is up and the wave must now topple over and crash. Quiet and still, it is frozen in that last instant before any white caps appear. Even today I wait for that wave to fall.

I know this eastwardly dipping ridge of rock well, the way one knows an old friend that you have seen and visited through many seasons. And it is indeed an old friend now; we are bonded by the shared years of our companionship. As I drive out in the early mornings, I see the highest parts of the Ridge catch the first light of the sun, rich in the colors of a morning spectrum displayed on what seems to be the largest drive-in theater in the world. I know those brittle rose-red colors of sunrise that change instant-by-instant in the first magical moments of the day. They paint first the Ridge top, and then race night's retreating shadows down the Ridge itself.

In its wintertime look, deep stillness becomes a blanket that wraps the rock, broken only by the occasional elk or deer standing casually up on its slopes or perhaps even down beside the road, panting a frosted breath. I also know the softer more inviting shades of summer green when the upper sharp rock edges are blurred by vegetation and I may see a fox or a coyote moving across the road in a search for the eternal next meal. Spring and fall balance these two—for they are similar seasons going in different directions.

I do not accept that spring is truly near until the deer appear along the base of the Ridge, drawn by tender shoots of green grass that satisfy a hunger born of winter roughage and abstinence. In the summer, the deer move hidden in cool shade among the small trees after first light, but then begin to appear more frequently on the open slopes as fall approaches, fat and enjoying the crisp bite of the air. It has changed little across the more than 20 years of seasons I have known the Ridge. Instead, it is I who has changed, my eyes carefully scanning a land that enfolds me in imagery both powerful and sublime. It is my own eyes that see a little deeper now and it is my thoughts that imagine their way into the rock, and into a distant past.

This Ridge like the Earth itself means different things to different people. In the summer it adapts comfortably to the feet and tires of hikers and mountain bikers, the bikers often rushing by, capturing only the shape of the land, the turns, and the downhill sections. The hikers breathe a little slower, and perhaps mentally a little deeper, yet many only know the feeling of comfortable shoes along the trails, alert for wildlife, the deer, or perhaps a rattlesnake resting in the shade of mid day, watching. But as they walk along Dinosaur Ridge, they are going back in time, way back, back before any human breath had yet breathed the air, before any human footprint left its trace upon the Earth. I wonder what the air was really like then, what smells did it bring, was the sky just so blue; was the shallow ocean green and clear? The Ridge calls in many languages and to a lover of the Earth it sings a particularly beautiful and rich music, for it is here that the stone pages of this ancient book of stories have come open, exposing their secrets.

The river sands, the ancient beach and the footprints remain but the dinosaurs and the ocean have long since gone. Plant and animal fossil remains and ripple marks are the only reminders of the storms and the waves that once defined this old coast. It was only about 130 years ago that some of the very first discoveries of the giant dinosaurs were made in this area, including Stegosaurus.

What is there about this Ridge that calls to me, for it is more than dipping sandstone and clay. Today it marks a boundary line between the mountains and plains as sharp as if drawn by a great artisan using a very fine brush. This area, this boundary has earned an illusion of permanence, a rock anchor through tides of change. One hundred million years ago this area also formed a boundary, then between the land and a shallow sea that extended some 800 miles to the east. In those times dinosaurs roamed and migrated along that ancient beach at least from what is now Boulder, Colorado in the north to New Mexico in the south. One of those dinosaurs was a hollow-boned ostrich-like critter with the budding characteristics of modern birds. Today some birds fly along this very same pathway in their own migration high above this ancient beach, their feet only touching at night to rest or to eat.

The Ute, Arapahoe and Cheyenne saw this place without the roads and the houses, and they too were touched, connected in a very real way with the land that supported them. In the rock we see lessons from the past and with an inner ear we hear voices of the Earth. We listen in the wind for whispers of a message of connectedness that draws together all the loose ends of past and present like some embracing tapestry of time.

What are these magical and mystical connections with the Earth; why is it that people of all kinds are drawn back to her in wonder, even as we and all other life have been drawn from her? We hear the call of nature but are sometimes deaf to its richness. Nature is not just the life itself but all that adds sustenance: the rivers and oceans; the air with its clouds and storms...the mountains with their volcanoes that give them heat and heart.

There is a modern-day drag strip that runs along the base of the Ridge just south of here. In earlier times it was an automobile test track. It strikes me that this area has always been a test track, a place where the species *du jour* is tested to determine its survival, no esoteric discussions of rights and privileges, just actions in the zone of sacred balance between the eater and the eaten. Many who came before do not remain. Humankind is but the latest species in this endless test of survival. We hear that the Earth is sick and I wince a little at the sum of our collective audacity. The delicate balance of Earth's systems is certainly stressed by constant and sometimes dramatic change and as a species we have indeed been thoughtless and negligent in our caretaking. But it is not the Earth that is in jeopardy, rather it is we who live on its surface, play in its rivers and ride its currents of air.

It was the Earth together with Spirit that brought forth its children in all the splendor and wonder that we see. All life forms are children of the Earth and whether humans survive or not, the Earth will continue to bring forth new children in its own season, in its own time and place, and of its own form. The only real question is whether we will be around to see it.

Away from this Ridge in homes around the world mothers and fathers are teaching their children every day about their deepest beliefs. Sometimes they teach in words, but always in their actions. From the beliefs that children accept, they form values about how to live, about honesty, cooperation, a work ethic and what is good and what is bad. The Earth, mother of all life that we know, teaches her own values if we can but listen. For the great act of creation that formed the Earth and the stars and even my little Dinosaur Ridge is not over, not even close. It is an on-going process of infinite proportions. There are now millions of years of history reflected in the lives of the living and in the old survival stories of the long dead. The lessons we find in nature fit nicely in our own lives, and they can be used in our decision-making process to enhance our life experience. In that sense the Earth is a living Bible, its messages told in the processes that formed its mountains and in the life and death of its inhabitants, all played out over millions of years.

When we look beneath the first layer of the Earth's magnificence we can pull out common threads of connectedness amid patterns of change that are repeated over and over. From these lessons we discover values that the Earth supports, those practical functionalities that can be sustained on our planet and in our lives.

In Earth's wonder there is also an invitation to a journey of self-exploration for we are all the product of a physical and a spiritual journey, both important and both intimately linked to our experience on this Earth. We live entwined in relationships that form an immaculate and fluid balance, each piece incomplete unless acting with and against the others. Yet our most common and deeply embedded perception is one of separation; from the Earth we walk on, from the plants and other animals that call it their home and from each other. Our thoughts are framed in a self-imposed and illusionary island of separation consciousness and we have become masters of a world defined by "Us" and "Them." Our sense of separation and the enormous power of our perceptions have shaped our view of the physical world and impacted our spiritual evolution.

Thoughout this book, we will be looking at connections, cause and effect, and relationships, with Earth and with each other. We'll take a look at the Earth and what happens when things are out of balance, and finally we will explore values the Earth truly supports, analogies for our own lives and growth. At the end we may see ourselves in a different way and look again to the community that is Planet Earth.

CHAPTER 1

THE MANY VIEWS OF EARTH

We are all a part of a Universe of consciousness, a Universe of vast distances and enormous beauty and of strange and wondrous tales. We are an active part of that Universe, as much a part of the Earth as the mountains and the oceans.

It is easy now to close your eyes and picture Earth from the distance of space. It was not always so. For most of our history we have gained the distance to look back upon the Earth only through our imagination. But today it is the sparkling detail of photographs from satellites and the Space Shuttle that provide our imagery. There it sits out in space, a beautiful and magnificent blue sphere, speckled with clouds, ornately laced with oceans and continents–solid, liquid, and gas, the three phases of matter. And regardless from which angle you look, it presents an almost perfect circle to the eye. After seeing thousands of photographs we still pause, captured by its beauty and magnificence. Remember this image; we will call upon it again.

From the perspective of deep space the Earth would seem to have been thrown rather carelessly into this dark corner of the Universe, a speck of coagulated dust in a bleak neighborhood of a few planets revolving around an undistinguished, middle-sized and middle-aged sun. The Earth is after all, a mere 8,000 miles through its center in a Universe that stretches across billions of light years, most of it empty, broken only by scattered brilliant points of light. Still, it contains the sum of our history and the story of our struggle for survival, and then for enlightenment.

In the shelter of our homes, schools and boardrooms, we sometimes see models of the Earth hanging from the ceiling, suspended by strings. But in the deep truth of space there are no strings, no supports, the Earth doesn't really hang; it simply *is*. And it is largely what it appears to be: a beautiful and complete dynamic system—our home.

Looking again from our view in space, we see a quiet and unbroken planet with no apparent activity except the slow rotation that separates night from day. The moon lingers near in a close and serene orbit, its face always to the Earth as if in witness to an eternal quiet. But the history of the Earth has been far from quiet and the Earth is not without motion, and it is elaborately far from being a simple place. Instead, Earth displays a deep restlessness marked by wonder and struggle, abundance and starvation, beauty and violence. It contains all opposites within its sphere and like a painting we are able to view it in many lights. Ultimately we find and focus on that special facet that resonates within each of us.

For some it is a platform from which to view the stars and ultimately stage a journey to make their acquaintance. For others, the Earth is an orbiting mine for gold and other precious minerals, best appreciated when laid open and vulnerable. In this view it is a container for the precious, but not precious in and of itself, a thing to be gutted and often left un-reclaimed. For millions more, Earth is a garden of seemingly unlimited proportions, its soil continuing to act as the perfect medium to bring forth our daily nourishment. For still others it is a recreational planet of tremendous beauty. Its mountains and rivers provide countless playgrounds for all who come, offering nourishment for the body and the spirit.

For me it is the imposing quiet of the Colorado high country sprinkled with distant sounds of birds and the whispers of winds through the aspens that calls me to silent attention and introspection. These sounds mark the beginning of spring in my little piece of the mountains; the raven and hawk, the elk, and the aspens, pine and fur all seem poised, waiting as the Earth prepares to pull back the heavy blankets of winter. The first camping trip of early summer always begins with an odd reconnection. For there is one object I leave hidden among the trees each year after summer's last camping trip, a well-worn outdoor chair. It is a comforting relic that dwells in silent witness through the long winter that I am not gone forever, that I will return.

In years past my Samoyed companion would lie in the thin shade of partially unfolded leaves, and she too would pause, ears front, holding her breath for a moment to listen more closely. We would both stand on an open glade and stare at the peaks around us heavy with snow, watchful and imposing. The glaciers that raged here for thousands of years have long since gone, but reminders of their passing remain in the sculptured bowls that soften the once sharp peaks of nearby mountains. Across the valley herds of elk, lean and strained by the winter, mingle freely with the horses and cows in search of first grass. This is my place, known to me and me to it; I am the latest human to occupy it for a time.

There are many other beautiful places on the Earth and none more captivating than where the three phases of matter, solid, liquid and gas touch in a wondrous display of inter-connectedness. Do you know of such places? They are the seashores that drape the edges of our continents and islands around the globe. Here the ocean reaches gently up in rhythm with the comings and goings of our legless companion, the moon.

The turbulence of the waves provides mixing of nutrients for creatures of the sea and the land. The sun's light provides the energy for reef development and sustains a rich and varied community of life just beneath the ocean's surface. Differential heating produces strong air currents for birds, creatures of the air that nest upon the land but feed off the sea. At mid day the sun peaks intensely in bright yellow and then drifts unhurriedly through the afternoon to an explosion of reds and oranges at sunset. In the fading light the water changes from clear and inviting to dark and mysterious as if to mark a solemn passage in the sun's daily farewell.

We have always been drawn to such places and we often stand in line with other Earth inhabitants, for creatures of the land, sea and air all banquet at the boundary of the ocean's edge. Though we have populated the globe, many linger near the shore as if we somehow know from where we came. It is no coincidence that people are drawn to the ocean's edge and the restful sound of the waves. We sit there apparently daydreaming but we are in reality re-tuning to the subtle and seemingly endless heartbeat of the Earth. And actually no audible heartbeat is needed, for we get the same feeling when we contemplate the beauty of the mountains. As long as we are still and open to nature we can be re-tuned. It is nothing less than an unrecognized appreciation for the very deep and beautiful concert of the Universe in which we are all invited to play.

From rain forests and deserts to mountain peaks and ocean depths, what could be more different? Yet they are all made of relatively few different atoms arranged in a limited number of ways. And so at the atomic level we may wonder what could be more the same?

THIS MODERN WORLD

Many of us live with a nagging and unsettling feeling that life and we in it are just moving too fast. We are experiencing the classic illusion of the passenger on a train who passes a picket fence. The pickets go by so quickly that the fence appears to be solid, but upon closer inspection we find that it is full of open and empty spaces, much like our very own lives. Conversely, many others seem to focus only on the individual pickets, they do not always recognize the connection of each to the other and do not grasp the whole view, the fence. We may feel our lives too are full of disconnected events, struggles and searches, each without context and without a binding thread of continuity. We have discovered the map of the human genome but found it does not lead to our own happiness. Distraction can be a welcomed relief and the more we feast on the exhilarating wonder in this world the more we find to explore and bring into our treasure of experiences. Yet in the race to enjoy the magnificence of the meal, we consume too fast to saver the taste. We may miss the slow and delicate beauty in our own journey of life.

Most days we move too rapidly to even notice the land; we race across the Earth from place to place so fast that our feet do not seem to touch the ground and there is little contemplation of our "being-ness." We do not feel the Earth as a presence and we do not recognize our relationship to the rest of our world. This was not always so. For our ancestors, there was an important time to ponder in the waiting for the fish to bite, or the game to appear, time for contemplation and inner exploration. In their love of the mountains and the sea our ancestors were expressing an attachment to things that are timeless, attachments to a flow beneath the surface in our river of events. This undercurrent flows to the rhythm of the Universe, the protective certainty of the long term.

As a species we attack and we acquire, focused on what we believe we need with no time and little regard for things we do not see as immediately useful or necessary. The results have not all been bad, but there is a cost to our wholeness.

We have become a species deeply skilled in a few facets of our existence, but we lack the understanding and the strength that comes from our forgotten roots.

Our relationships come and go, far too quickly and often without depth. Surrounded by acquaintances we are looking for friends and we are lonely and often alone. From its beginnings the Earth has been a place of connections and relationships; and in a communion with its cycles and seasons there is an opportunity to re-ignite our own sense of place and purpose.

CHAPTER 2

WHY LISTEN TO THE EARTH?

When there is sadness and the heart breaks open
There is only the clear stream pouring
over stones to wash away your tears
There is only the sunrise, pure and clean to refresh you
There is only the moon
and the stars to make gentle your path
And there is the lightning to remind you
of the power of Earth

It is not surprising that many have come to feel their lives are significantly out of balance. But the Earth has been teaching lessons for millions of years on how to stay in balance and what happens when we get out of balance, if we choose to listen. Often, when we look closely at the Earth, it seems to be looking right back at us, so many cycles, relationships, struggles, successes and failures that seem to mirror in some ways the things going on in our own lives. And we may begin to recognize in nature processes for creating, and for sustaining and restoring, that seem to resonate with us and we may even begin to wonder: who's mirroring whom?

Still it is hard to value something so close you can always find it simply by looking down. What does the Earth have to do with it, anyway? Most were never taught to listen closely to the Earth and the idea just didn't register with many of us on our own. In fact, from an early age most of us learned to simply ignore the Earth. After all, it was always there in some shape or other, reduced to a fuzzy and unimposing backdrop to the consuming drama of *our* lives. We walk on it, we drive on it, we build our homes on it, we plant our crops in it and gather our fruits from it. We look at it on vacations and occasionally marvel at the power in it, and ultimately many are buried in it.

Most don't really doubt that there are a few lessons of survival to be learned from the Earth, but ours is often the low view of immediate needs, too limited to capture the breathtaking spectacle that is all around us. The depth of those lessons, the similarities in our lives to the processes of nature, and the raw wonder are too often missed. For the Earth is a living and immaculately responding system. Stirred into form among the stars, it continues to change and return to balance in a wildly imaginative co-creation with life. Library Earth carries the records of past events that have shaped our world and provide keys to navigating our future—not just the infrequent events we consider extraordinary, but here in our day-to-day lives.

I heard a beautiful concept long ago that goes something like this. We have all known people who acted as a "rememberer" for us in our lives. They were the people who appeared in difficult times when we were troubled and had lost sight of our true selves, our strengths and virtues and our potential. That "rememberer" may have been the parent who picked us up when we fell and helped us to try again when we were ready to quit. Or it may have been a teacher who didn't give up on us, a spiritual guide perhaps, or a close friend who was there for us in our time of struggle. Those people remembered the truth about us until we regained our perspective. And in so doing they helped us to put ourselves together again, to literally "re-member" ourselves, reclaim our potential and find our path again.

On a larger scale the Earth acts as a "**Cosmic Rememberer.**" It holds its own values within its creative process and applies them in its unfolding. The lessons of simplicity, harmony, connectedness, balance, and community that we find in nature are not accidental and they can be transcribed to our own lives with the certainty that this is the way the Earth works, and we can enrich our lives. It holds both our history and the keys to living well today. Even its vibrant and necessary seasons speak of familiar cycles of change that resonate in our lives. There is the hope of spring, a promise amid uncertainty. Summer speaks of purpose and continued growth, followed always by an autumn of abundance and harvest. Winter's time is contemplation and the sense of a yearly death as a stage for rebirth. One revolution around the sun—one cycle completed. The Earth has seen many cycles, and with them the birth and death of its mountains and oceans, and the rejuvenation of life after catastrophe. It reminds us of the interconnectedness of all things in the layered history of stone that captured and preserved the story of the beginnings of life, and we play them back in the context of today.

THE IMPORTANT ROLE OF LIFE

We know that there is nothing terribly unique about the chemical components that make up our bodies. In fact you can get a pretty good recipe for the physical making of a human being from a list of the eight most commonly occurring elements in the crust of the Earth with a few rare and important elements tossed in. Everything that is in humankind is found naturally in the Earth. In painting its great canvas of creation, the Earth used remarkably few colors, but many different brushes. We belong here and in a very real sense we are as much a part of the Earth as the mountains and the oceans—we are of the same fabric. In this recognition we uncover the first layer of our connectedness and look more closely at a Mother Earth that seems to stare back, patiently waiting for her children to learn to communicate with her.

Life, through the ages and through its own changes has had much to do with the Earth becoming the spectacular planet it is today. We did not evolve on our own but in a close partnership with the Earth itself, for both have been changed by the other, both have evolved and unfolded in a long and intricate dance of give and take, stimulus and response that questions what is life and what is consciousness. Do we carry the unspoken answer in the wisdom of our genes that reproduce organisms blue-printed to lie in close alignment with a sustaining Earth? Could it really be that the evolution of life and the evolution of the planet are in some sort of loose-knit partnership of co-creation? For life did not come all at once and it did not come into an already formed and perfectly suited environment. When the first reproducing bacteria appeared a few billion years ago the Earth was a hot, smelly and frightful place to be.

Just for a reality check let's take a moment and hop into our mental time machine and travel back about 450 million years or so. Considering the Earth is about 4.5 billion years old this is really just a hop, skip, and a jump back in time. By then the engine of life was well under way and the oceans were populated with a variety of critters both large and small including relatives of most of the ocean life we see today.

Ancient reefs were widespread and waves crashed upon the shore, sometimes roughly against the edges of mountains and sometimes more gently, breaching on soft sand beaches that the wave action had helped to create.

But beyond the shore of a rich ocean stretched the land—what of the land? Would things really look very different than they do today? Well, the land was certainly in different places about 450 million years ago, most of the major land masses were below the equator. But the same physical processes were at play, rains swept the Earth and streams worked both high and low to wear the mountains down and carry their loads to the sea.

What else would we see? Well, it might be easier to start with what wasn't there, for the land was barren. In those times there were no land plants, no beautiful trees of oak and pine, no redwood and fir, no sweetgum and magnolia, nothing. There would be nothing to block the wind, whether raw cold or scorching hot, except the bare rock. There were no shrubs, no grasses no corn or other grain, and of course, no flowers to break the monotony of the dust, rock and dirt. Perhaps the flowers would not be missed since there were no insects to be attracted to the flowers and pollinate them. There were no birds to eat the insects that weren't there, no animals of any kind. But then again there were no roaches and no mosquitoes, either.

What have we missed? Oh yes the smells, what of the smells? We could sniff for a long time without picking up anything that reminded us of home. The variety of smells we capture everyday without thinking about it would be notably absent, no rich smell of soil, tree or flower. But there would have been the acrid smell of sulfur and some noxious volatile gases if you were in the right spot.

On a clear day we might feel the sun's heat on our backs—strong, bright, and intense. We might wish we had taken a hat because there was no shade, simply no shade at all from the mid day sun. Can you really imagine that? Looking at the ground more closely we see that the land is either rock or dirt, but no soil, nothing that would even remotely welcome the seeds we know, if indeed there had been any. In some areas mud, lots of mud from the rain eventually became baked into a dry and unforgiving hard pan. Looking up for our familiar deep blue sky we might find instead a sky of sharper contrasts and paler colors. In those times oxygen was still building in the atmosphere and there was insufficient ozone to block the deadly ultraviolet rays that blanketed the Earth in its own efficient form of sterility.

As we speak to each other we may be startled at our own sound, perhaps the only sound to break the staggering weight of quiet. For away from the oceans and the streams, the only sounds would be the wind, occasional rain falling on a rock, lightning, a distant volcano, perhaps, and maybe the occasional landslide. It is almost impossible to imagine that kind of quiet, pervasive and interminable. Yes, in walking the land from shore to shore on any continent, what we would find is quiet, lots and lots of quiet, almost frighteningly quiet, and of course plenty of mud for everyone.

How Did the Land Get From Barren To Beautiful?

The long progressive journey of life, failing countless times only to spring forth again, left in its wake much of the beauty we see today. One organism impacts little, but together and interconnected, life built upon the sum of its participants with miraculous results. The by-products of life processes that began billions of years ago have shaped the

Earth as much as any change that came before or since. Instead of being given a ready-made and friendly neighborhood, life shaped its own environment with the tools at its disposal in an unprecedented display of nest building on a planetary scale.

That blue-green algae we see so often on the walls of our aquariums is not just an irritatingly persistent neighbor, it was a key building block for our atmosphere long ago. Blue-green algae are bacteria that hold the record as the oldest known fossils of about 3.5 billion years and if they could survive in that ancient and unforgiving world you begin to see that overgrowing your aquarium walls in the steady comfort of your house is really no problem at all. No wonder algae are so hard to get rid of.

Gathering energy from the sun these very first bacteria were anaerobic, consuming carbon dioxide and releasing oxygen as they began forming our atmosphere.

And it was here in the atmosphere that ultraviolet light, striking the occasional oxygen molecule provided sufficient energy to create another form of oxygen, ozone, which in turn dampened additional ultraviolet light. Gradually, both the atmospheric free oxygen and our protective ozone layer were increased and the stage was set for life on land. For only then could plants and then animals safely leave the waters and climb upon land with a steady foothold.

The enormous grasslands and immense forests that followed from the first successful efforts to spring upon the land eventually produced the sustaining soil of this planet, a soil that we are now depleting at an alarming rate. This miraculous substance that has been nurtured and drawn from rock and sediments and molded over time by organic life, provides a medium for an unbroken chain of food that nourishes us. It has been described as Earth's placenta and truly, from it comes a harvest of births from around the world.

LIFE'S LONG AND WINDING ROAD

It was in school that I first heard the word "evolution," a term to describe the ongoing unfolding of life. Such an appropriate word, its very sound implies a steady gait on an almost inconceivably long and drawn-out journey that led all of Earth's creatures to this moment in history. Reflecting on the diversity of life both past and present, and the countless failed attempts of long lost species to gain a permanent foothold on the Earth, I wondered if this process could adequately accomplish the task. But I remember hearing that Johannes Kepler, famous for discovering the three laws of planetary motion, had also proposed the notion that "anything that can happen must happen over time" and time, the sculptor in Darwin's process, had billions of years to work with.

Planet Earth is never idle and during all the struggles, the failed attempts, and the blooming new life, the Earth's plates were shuffling, splitting and coalescing again and again over hundreds of millions of years. Meanwhile members of the same species that became separated as plates broke apart began to display different characteristics. Monkeys in South America display prehensile tails, the ability to use their tails as another hand to grasp and hold themselves in the trees. Their African relatives do not share this ability because the characteristic was acquired after the split between Africa and South America. For them, evolution's boat had already left the harbor.

Yes, change is an ongoing and permanent element of our history. And regardless of whether through billions of years of trial and error or through Divine instruction, we are here now, surrounded by this wonderful and elaborate ongoing experiment we call Earth. And we carry in our

DNA the recorded instruction of how to participate in the experiment. The Earth, while evolving itself, also nurtures those ideas that work in life's evolution and tosses out those that don't. The ongoing and self-correcting creation process that we see today brought this particular Earth community into being.

HOW FAR HUMANKIND HAS COME IN ITS DEVELOPMENT

Now we are at a place where our species by itself can affect significant change across the entire planet; that should make us cautious and thoughtful, after all it is our own home.

Our impact reaches beyond our clean world of steel and video games into shrinking rain forests, and increasingly polluted water and air.

I live near a place that was once home to the Rocky Flats nuclear weapons plant. Despite the harshness of its name it is actually a strikingly beautiful place set on gentle plains that drape the nearby mountains. Unmolested deer wander its flowing grasslands and sharp valleys giving an impression of assured harmony. The oversized streams that scoured the valleys once flowed furiously out of the mountains and owed their strength to the heady waters of melting glaciers. They were like streams on steroids and they changed the landscape. And like the glaciers, they are now gone. Looking away from the mountains there are views that stretch to an eastern horizon. The plains where the Cheyenne and the Arapahoe once camped are now richly dotted by more permanent houses and other buildings. It only takes a few pausing moments to sense that this is a powerful place. But there was a time not long ago when it produced power of another kind for different purposes. Rocky Flats produced plutonium triggers for nuclear weapons, efficiently but not without byproducts and scattered waste and now a tremendous effort has been expended to clean and restore the land. Ultimately it may become a wildlife refuge that allows human visitors. Today portions of it are largely sealed away and unavailable, waiting for time and nature to soften and distribute the isotopes and chemicals that humans created and brought together.

I remember the expression, "within arm's reach" and realize that as a species we have developed very long arms, for today there is little that isn't within our reach and experience. If we as individuals haven't traveled the world, many of the things we use and consume certainly have and I seldom meet a meal that has not traveled far to get to my table. The November 21, 2002 Environment News Service (www.ens-newswire.com) referenced a report from the Worldwatch Institute noting that some of the food we eat can travel "between 1,500 and 2,500 miles from farm to table."

How different this is from only a few hundred years ago when most people lived within eyesight of the fields and forests that supported them. Unless we live exclusively out of our gardens our food resources have come from some other place. This was not immediately a problem; the extent of the Earth exceeded our grasp, now it does not. As our cities sprang up they pulled in resources from further and further away on the increasing ability of our transportation system. But the point is there is a growing separation between the sources and destinations of our food.

Old wisdom says that the further we are from our source the more vulnerable we become and in truth it is unlikely that any large city could survive for 30 days without re-supply.

THE PARADOX

The world has progressed in amazing fashion but not evenly and our generation was born directly into the split between poverty and panache, between starvation and lavish wealth. We imagine the wondrous examples of our technology, the skyscrapers, the bridges, marching into the future seemingly without end. Truly what seeps into the consciousness of humankind can be brought into reality. But what of the rest of the world? Remember the spectacular image of the Earth from space that you held in your mind, the one you created from your memory? It may have been a composite of all the snapshots you have ever seen of Earth, or it may have been a special one. Snapshots are like paintings and we can see many different things in them depending upon our mental perspective. As we go back and look deeply into our snapshot of the Earth we might be reminded that of the 6.28 billion people on the Earth today, nearly one half make less than $2 a day according to the Worldwatch Institute report, *Vital Signs 2003*. Poverty's tale is written in insecurity, inequality, poor health, and illiteracy. We have conquered polio, small pox and many other global diseases, yet according to the World Health Organization, *Africa Malaria Report 2006;* preventable malaria still kills an estimated 2,000 African children each day. Do we lack commitment to the cause? Poverty and disease together with environmental degradation lead to instability and war.

Our glass and steel structures that rise above the Earth are actually sustained by the resources that supply them and we see that we are not as stable, not as indestructible as the outward image of our buildings appears.

BIG CHANGES

There are big changes afoot on the planet and we can see these too in our snapshot if we look closely and think deeply. Not that change is new; it is instead the most common thread connecting past to future. But some changes we see imply a global climate in transition and come at a time when we are more and more aware of our limited resources. These changes will require us to test our values against all that we have learned about the fragile connectedness of all things.

If we look up from our coffee and breakfast on any given day we are inundated with information that tells us the Earth is changing in subtle and sometimes dramatic ways. Make no mistake—the Earth remains an abundant and prolific source of all we need. Yet in the news we see disquieting undertones. Between the headlines of war, greed and personal tragedy we can hear a distant rumble like a far-away locomotive in the night, and we know instinctively that the tracks run very near our own door.

One story tells that California's largest grocery retailers began displaying signs cautioning customers of the dangers of mercury in fish. For a time, they advised not to eat swordfish and shark, and to limit consumption of fresh tuna. The April 11, 2003 Environment News Service reported that California filed lawsuits against 18 restaurants for not warning customers of the risks of mercury in seafood. Who can remember grocery stores and restaurants warning their patrons of the food they are supplying?

From fish and meats to fruits and vegetables we are more and more wary of the quality of our food. Few of us have ever come into contact with enough mercury at any one time to be concerned. The problem is that organisms cannot easily excrete mercury and so it accumulates and causes problems all through the food chain.

I remember the experience of having a whole jar of mercury to work with in high school chemistry, of putting in a dime and watching it float on the surface, of seeing the marvelous shine it produced. It seemed a wondrous and beautiful thing and indeed it was. Of course I didn't realize at the time that it was toxic or that it has the ability to be absorbed directly through the skin. Mercury was once thought to be rare, but we don't need to go out and find mercury, now it comes to us in our food. The April 24, 2003 Environment News Service cited state official guidelines for consuming fish which pointed out that now "all fish, whether store bought or sport caught have some mercury".

GLOBAL CLIMATE CHANGE

Global warming is the latest change that has captured the world's attention. Climate change was not something anyone really thought a great deal about when I was growing up. Conflicts with the Soviet Union, landing a man on the moon--these were the things that held our undivided attention. I just don't remember the role of day-to-day weather ever getting much credit as having been important in the history of modern man.

But climates do change; they have always changed, sometimes slowly, sometimes dramatically. And the extremes in our weather have affected the outcome of some of our most significant historical events. The wins and losses, the great achievements or failures were usually laid at humankind's door for pride or criticism but we have now developed a greater understanding of the impact of weather on our history.

Historical mysteries such as the disappearance of Native Americans from Mesa Verde and sites throughout South America as well as the Vikings now seem to be the result of one or another form of climate change. From the defeat of the Spanish Armada to the mini-ice age which ran from as early as 1150 until about 1850 and ushered in a period of crime, disease and mass deaths, weather and climate have always been important factors profoundly affecting our history and our development. ("The Little Ice Age in Europe" by Scott A Mandia, www2.sunysuffolk.edu)

We did not find the hand of man in these events, and we know that the Earth's climate changes dramatically on its own. But now we are at a place where our actions do impact our climate. And nowhere is climate change more visible than in the rapid melting of glaciers and long-frozen ice fields around the world. From the Arctic down through North and South America to the Antarctic at the bottom of the world we see the same pattern of glacial retreat. Is there a cause for this change to which our names are attached or is it just the product of periodic climatic change? Only now are we beginning to understand Earth's balancing processes sufficiently to say that yes, the activity of humankind is a significant, though not the only cause.

For over 30 years constant calving in Antarctica from the Ross Sea to the Antarctic Peninsula has reduced the ice shelf to a size not seen for some 12,000 years. If trends continue Arctic Sea ice may be gone entirely by the end of this century. In Africa, Mt. Kilimanjaro's glaciers have shrunk almost 80 percent since 1912. In the Southern Ocean, the stories of change are spectacular, even if the witnesses are few. In recent years, huge icebergs, some the size of small states like Connecticut or Delaware, have broken away from the Antarctic ice shelf and moved north on prevailing ocean currents. These icebergs disrupt the Antarctic ecosystem but should this really matter to us?

I grew up in Alabama and we didn't know much about snow but still I marveled at the thought of polar ice caps. Those far away and vast areas were fertile grounds for pure imagination. I could not have envisioned anything more different from my place, my world. It seemed unbelievably clear, cold and clean. Yet the Arctic and Antarctic ice shelves are not just some frozen chunks of ice stuck at the ends of the world; they play a necessary role in stocking nature's grocery store.

Both poles develop strong bottom currents of dense highly saline waters often amazingly rich in nutrients. As these waters spread away from the poles they mix and collide with other currents moving upward and bringing nutrients near the surface where they drive the productivity of the world's oceans. The location and activity of these converging and diverging currents hold the key to the balance of nutrient mixing in Earth's oceans and to maintaining the long chain of life at the end of which we find ourselves.

And so we learn that the story of Earth's oceans is not complete in the tales of trade winds and surface currents, the ones depicted by little blue arrows on many of our world maps. We see again an intricate balance and a subtlety in nature, an importance of those things below the surface. And we also find that something little discussed and far away has a large finger in our food supply.

Do we begin to see that the lesson in connectedness is far-reaching?

At the same time we see changes far away we are also continuing to learn that our natural habitats are more connected and more fragile than we had thought. The melting glaciers are stunning in their possible impacts, but strong climatic change occurs periodically in nature. Perhaps these potentially catastrophic changes were already incubated but are now spurred on by humankind at a far more dangerous pace. There are other changes we see in our snapshot that even more directly result from the hand of man.

WHAT HAPPENED
TO ALL THE FISH?

The May 14, 2003 Environment News Service cited a long-term Canadian and German Study that "ninety percent of all large fish in the world's oceans are gone and just 10 percent remain after commercial fishing vessels have taken their tolls over the past 50 years," both warm and coldwater fish alike. That includes Marlin, Blue-fin Tuna and even Antarctic Cod. And so the top of the ocean's food chain is in danger of being lost. We really do not know the full consequences. Will we be the species remembered for literally eating out the top of the ocean's food chain?

In my own lifetime I have seen and noted the changes. When I was very young I had the chance to go deep sea fishing out of Destin, Florida, a favorite place along the gulf coast. It was hardly sport; we boated out about 20-30 miles, used sonar to find a school of fish in about 110 feet of water, put three hooks on each line and dropped it over the side. The only real challenge was to get the line to the bottom where the Red Snapper were before some other fish took the skipjack bait on the way down. Occasionally we pulled up a Grouper and stunningly in hindsight we threw it back. It was considered a junk fish and we only kept Red Snapper. Can you imagine throwing a large Grouper back today? Years later, I learned that red snapper were getting harder and harder to find, and lo and behold, restaurants began to promote grouper. Eventually the Patagonian Tooth fish was marketed as the "Sea Bass" which sounds so much more delicious, but in those times no restaurant I knew of sold Sea Bass. Now Grouper are harder to find and even Sea Bass are feeling the pressure. I remember discovering a few years ago that it was hard to get Cod at my local supermarket; it was replaced with Haddock or Pollock or some unknown variety of other whitefish which had yet to be given a tasty name. But where in nature's womb is the next white fish? So, yes, even I have noted this decline in fish and wonder at the eventual outcome. Sometimes we hear that perhaps the best we can hope for now is some sort of limited ocean restoration augmented by a high degree of farm raised fish. But that does not take us completely out of the woods. While oceanic fish may contain high amounts of mercury, farm raised fish often contain elevated levels of PCBs.

There is one particular place in the oceans where fish gather in a remarkable abundance and variety: the living reefs. Some species call the reef home while others only visit to play or to mate. Still others come to hunt while some come to hide in the protection of this intricate and massive mixture of living tissue and secreted stone. Forget about humankind and our achievements for a moment, reefs are the largest structures on planet Earth made by living things. The Great Barrier Reef alone extends some 1,500 miles along Australia and can be seen directly with the naked eye from space. But wait—there seem to be changes in our reefs as well. In the Caribbean rising water temperatures, increased population with its developments and resulting increased sediment, chemical pollution, and overuse have all contributed to a significant reduction in coral. Adjacent to the Caribbean we find the Gulf of Mexico where restricted circulation causes it to act as a giant bathtub collecting the sediment and the pollution from every river that enters, and there are many. Globally we may begin to wonder how many can be put in Earth's bathtub before no one gets clean.

No living organism, man included, lives in a perfectly stress-free environment. Instead, all creatures live with the collected burdens of minor disease and accumulating toxins. Generally we can still survive, but our resistance and our ability to recover from illness decreases as our age and our burdens increase. Coral in the Caribbean has its own burdens to carry including the occasional hurricane whose waves may break and damage the reef. Normally it can recover from mild climatic change and storms but when the accumulated burden is too great, it too dies. In this case we have the clear knowledge of what to do, of how to limit our sediment and chemical pollution, and how to limit our overuse. It is the economic motivation and political will that we seem to lack.

As we move away from the oceans we find that the land is also bending under the weight of our needs. Great portions of the Amazon rain forest continue to fall to slash and burn agriculture despite greater awareness of the irreplaceable value of the forest to supply rare woods for our needs and medicinal cures for our list of pressing diseases. Around the world our rain forests, like our oceans, act to moderate climates and they supply free oxygen while removing carbon from our atmosphere. Although we have lessened our rate of deforestation of the Brazilian jungle, National Institute of Space Research data showed 4,621 square miles of rainforest lost in 2008. There will be no immediate replacement to this ecosystem that took millions of years to develop.

Let's look again at our snapshot of this beautiful planet, what do we see this time? Perhaps the first thing that strikes us is a lot of water, beautiful, blue and deep —but less and less to drink. The May 27, 2002 Peoples Daily Online cited a World Health Organization statement that, "At least 5,000 children die daily of diseases caused by consuming water and food polluted by bacteria." In the islands of the South Asian seas millions of people live their lives without basic sanitation. In the future we will desperately need clean water for our homes and cities and for growing our food. Without abundant clean water, the picture of our collective future becomes dim. Estimates suggest that the Earth will support a population of 9.2 billion by the year 2050, and we must wonder where they will get their water. Environmental degradation is just bad housekeeping. And its effects are mounting.

TRENDS

In the longer view, we are not concerned with one season, one year, or perhaps even 100 years. We look at trends, knowing that as the Earth maintains balance with the sun and its own shiftings, there will be movements of change back and forth, hot and cold, wet and dry. When we take the same approach to our own species we find a disturbing pattern: as we have grown and developed our capabilities, we have taken more things into our own hands. We have modernized our homes to produce a new environment separate from the climate outdoors and much more suited to our comfort; and we have discovered or invented those things that satisfy our physical needs and brought them to ourselves from around the world. Our achievements are strikingly magnificent and unprecedented for life on Earth. But we have also moved away from an understanding of our relationship to the Earth and to the other life it supports. We have treated the Earth as just another commodity not recognizing how interdependent we are, and we have found that nature is less tamed and under our control than we have thought.

We have created a distance in consciousness from our source.

Each day we awaken to the sunrise of an Earth moving in its own cycles of abundance and productivity. From accumulating toxins to global warming we hear messages in the responses of Earth that it is time to rethink our actions and align with nature's design. It is time to restore the balance in our thoughts and actions.

RETHINKING OUR RELATIONSHIP TO EARTH

Why listen to the Earth, why study the way mountains are made, the way rivers flow and sculpt the land, the way great storms destroy and cleanse, or the way other life adapts to all of Earth's changes?

The first reason is purely logical and follows from the fact that we are a part of the Earth, born of the same fire and forged in the same furnace.

The Earth has been around about 4.5 billion years. Life has been around about 4 billion years and it has moved and molded itself to lay in close alignment with Mother Earth. Have you ever heard the expression "Align with the Design"? There is a design for the Earth and when we live honoring that design, our lives work better, we are happier, there is more harmony, and we are in balance.

Earlier in its history humankind did not separate matter into the animate and the inanimate; they were seen as originating from the same source. In that context we are the Earth itself expressed in a different and self-replicating way.

As such we have experienced similar things, responded to the same changes. We begin to see that nature's macro-processes appear less random than we had thought, and we can see in Earth's long history reinforcement of those values that work and suppression of those that do not. Who does not gain in listening to an elder?

Life of which we are a part did not sit idly by and watch the Earth from a distance. Instead, life acted with the Earth in a co-creative process spanning billions of years and our DNA is keyed to living on the Earth, understanding its cycles, recognizing its creative processes. Life is not a Johnny-come-lately, but was around and a part of the unfolding process and it danced and dodged with Earth through all the changes, severe cold and heat, floods and droughts, the volcanoes, the meteor impacts—everything. The Earth was not a stage upon which life appeared; life was a participant in the construction of the theater.

Slowly we begin to understand that the process of life's development was never really separated from that of the Earth. There is a close kinship in the principles of creation, the principles that show us what works and what doesn't. The same processes that created and sustain the physical Earth, act in our own emotional and Spiritual lives.

The Earth has seen change upon change and countless failures that only represented steps to success. Through each change, the Earth has responded under guiding processes that moved to restore an old balance or to create a new one of different surviving components. Along with the violence and starvation we also find patience, harmony, abundance and always the unexpected; Earth never stops remolding today into tomorrow. We can see that our relationship to Earth is perhaps deeper and more entwined than we suspected and that there is much we can learn.

The second reason is that the Earth calls us to listen; the call flows from an ingrained and deeply seated connection that transcends our intellect and nests in our emotions.

In our quiet and thoughtful moments in nature we are responding not just to the beauty of Earth but to the consciousness that created it. In Earth's creating and sustaining,

Natural Processes act according to Universal Principles and these principles represent values the Earth supports.

We all listen on some level. We are listening when we are mesmerized by the rhythms of the waves at the beach and we are listening in the stillness of the mountains when we are awed with a sense of power. Every cell in our being vibrates in an envelope of Earth influence. We may not recognize it, we may suppress it, we may ignore it, or we may assign the feelings to something else, but we all listen.

In almost every life journey we come back at one time or another to nature for solitude, nourishment and healing. In the stressful times of our lives when all else is stripped away we find that a constant and supporting Earth and all its wonders are still there, waiting...unblinking. Our communion strengthens our attachments to things that are timeless, and those things center us in times of uncertainty. They touch the soul and the heart and our ebullience spills over in poetry and song.

The Earth is an ultimate sounding board that echoes our deepest mysteries, a full partner in our own evolution and to a real extent—we in its.

We carry the history of the Earth in our DNA and perhaps in our souls. When we see the Earth in this fashion we begin to understand our place in this drama.

CHAPTER 3

WHAT DO YOU MEAN I
HAVE SEPARATION CONSCIOUSNESS?

The Earth accepts and enfolds all peoples, all animals, all plants, all likes and all dislikes and all paths to enlightenment. In the time before we were able to vision other worlds, we believed that the earth was all there was. And in all that we saw and all that we touched there was no sense of separation.

It is our sense of separation, our belief that we are somehow completely different and separate than everything else that keeps us from perceiving our deep connection to the Earth. Many spend a life absorbed with this illusion of separation, and it is a view that ultimately makes of Earth just another thing, rather than a profound source of ancient wisdom. We speak of our lives as coming from the Earth and then returning to it, as if in the interim we are separate and able to live on our own without nourishment and support. The lessons we are taught as well as our own observations of here vs. there, up vs. down, dirt vs. flesh, have molded our perceptions with a deep and longstanding "awareness" of distinction, of separation. Yet there is a cycle, a rhythm in the Earth as strong and as striking as that of any human heart. Both beat in concert with the same constant vibration of the Universe. There is a deep significance in the thought that the Earth was created in the same process that created us, and that it is going through the same space-time experience. Both we and the Earth are active players expressing cause and effect, and endlessly dancing in and out of balance with ourselves and with each other.

A few years ago I was returning to Denver from a week-long business trip to Washington, DC followed by a very sad and hectic four days of working with my sister to clean out my parents' house. We had recently moved them into an assisted living facility and we were working to prepare their house for eventual sale. In the house we found the memorabilia of our connections, a reminder that in a very real sense we are the legacy of our parents, continuing an ageless journey through time. Looking at the old photographs, some of myself, I could see how I changed to look more and more like my father, a reflection of the DNA link that forever binds us. But just as important, we are also connected to our parents in our thoughts and emotions through our early experiences with them; I am still influenced by their unflinching perceptions of the world.

As I studied those pictures I had a strong sense of looking back through time. At one end was my father the strong athlete sure of himself and defining his place in the world; at the other end there was an old man bent by the years and using a walker. And I noticed that while the features of my father's face changed, wrinkled and sagged as he grew older the optimism for life in his eyes never did. The feeling of bridging across time followed me for several days.

Upon arriving at the Denver airport I took an escalator down to catch the airport train that connected the terminal to the various concourses and the baggage claim area. Aboard the train I found myself listening to the conversation of two women; one held a 3-month old baby girl and they were talking about her. As they talked and laughed the train passed through a tunnel decorated with eye-catching lights and moving figures that danced in the wind of the train as it went by. The baby looked for brief moments at the people around her but I noticed that her eyes were continually drawn to the light and the moving objects outside the train on the walls of the tunnel. She was not able to process the objects or to even understand the nature of the light but her eyes were large and round as she followed the wonder of it all. To her the objects must have appeared as an endless collection of independent and unconnected curiosities. She could not put the train in any context or yet understand the movement of her own journey.

Our sense of separation can cause us to see the world this way, as unconnected bits of wonder. But as we look closely at the things around us we discover that they are all made of the same kinds of atoms and molecules, basic building blocks that make up the erector sets of the Universe. In reality we find connections in the strangest places.

SORTING SKILLS

Early in the lives of children we help build their ego and their sense of separation by teaching them literally "to sort things out," to separate things. Children are asked to place square blocks in one pile and round balls in another, or they are asked to pick out the red or blue item from all the other colors. That's the process of sorting and we actually measure the intelligence of the child to some extent on their ability to separate one thing from another. The ability to separate and see differences is a basic tool of identification, for if we are to name things we must separate them from everything else. And it helps a child begin to understand the dimensions of their own body, their material identity, their relationship to the rest of the world. Skills of separation are highly valued in our society.

So how does this play out as an adult? Let me give you an example I did with a friend of mine—using a plastic knife and fork. I held them up for her to see, a knife and fork both made of the same black plastic. Like most of us her mind immediately recognized the different shapes and from that, what the tools are used for. Certainly as adults, we easily note their differences and no one would confuse the two of them. But if we look a little deeper at what they are made of, we see that on a chemical level they are identical. And if we look at their physical properties we would also find them identical.

Only their surface characteristics allow us to separate them so easily. They have so much more in common than they have differences. But we call one a knife and the other a fork based on their shape and purpose and set them aside in our minds and seldom think of them again. We do that with a lot of things—we see their differences and forget about the depth of their similarities. This creates enormous problems when we do this with people. I remember hearing a saying long ago that "to name something is to know it." I thought that was so wise when it was actually very simplistic. This is the trouble with being entirely comfortable with our sorting and our naming. As Eckhart Tolle paraphrased in his book, *A New Earth,* when we name something we tend to dismiss it and seldom look at it again in any deeper context.

As an extension of our sorting skills, we are taught early and often in this life to look at the world through a lens of separation: we identify differences—different people, different languages, different clothes and different cultures.

It is sadly unfortunate that we spend so very little time learning about the deep connections that link us undeniably as children of the Earth. These threads of similarity define the degree of our common humanity. It isn't surprising at all that most people see the world as they have been taught, through a lens of separation. We certainly can feel completely separated from others inside this individual body of ours, this body that sees colors and tastes things in its own individual way. And how can we be sure that others feel the same pain or pleasure that we do, see the same rainbow, feel the same joy of heart when we can only describe our experiences in words? Words are such poor vehicles, hardly adequate to transmit the depth and character of our individual sensations.

Many seem to live lives of disconnected split seconds, with no connecting continuity and often no sense of direction. In our darkest hours we can feel a most desperate loneliness and a sense of being cut off from the flow of friends and life. This is the illusion of separation and most of us have felt it at one time or another.

Native American peoples have lived close to the land for countless generations and many still do today. Unfortunately, in our western culture many have developed an increasing sense of separation from the land; our source for so many things. We gather our food from the land and the sea, whether plant or animal, however, with the conveniences of today we do most of the "hunting and harvesting" our ancestors did at the local grocery store. In purchasing our food many forget that our nourishment arrives with great effort and sacrifice.

I grew up on a farm and I remember years when we had a very large garden. When we wanted vegetables in the summer we often just needed to go out and pick them. But of course it is never quite that simple. I learned a little about the time and the effort it took to bring these vegetables to our table, the planting, fertilizing, watering, the weeding and the waiting, and finally the harvesting, peeling and cooking. And each year we tried to take and butcher a steer about two years in age. Because of the experience of taking care of cattle and occasionally watching a steer being butchered, I know much more about their lives and their struggles and what a steer looks like on the inside as well as on the outside, what it smells like and what it feels like. And I better appreciate the time and effort in the birthing, the raising and ultimately the sacrifice of the steer with its life so that we could eat. And it was actually much more than that. Working the soil in early spring through late summer for our crops, and feeding, protecting and caring for cattle through the cold winter and early spring when calves are born, were more than physical actions and they provided for more than our physical needs. The experiences brought satisfaction and fulfillment to the soul, a nurturing, contentment and a peace of mind.

These important lessons are lacking for many children today, those who learn to eat but not to value the time and effort and the sacrifice. It makes one wonder how the next generation will view their relationship to the Earth. Our technological advances have brought us incredible convenience and a steady stream of products, yet we make a mistake when we lose the sense of our physical and emotional connection with nature.

Science looks at connections and how things are related to solve problems and understand complex systems. In this way it helps bring us back to a recognition of our own connectedness and through that a compelling sense of our unity with the Earth and with each other. Scientists are very used to studying small pieces of things and then extrapolating that information to an understanding of a much bigger picture. Simply put, science often grows by *seeing the piece and imagining the whole*. The problem comes when people see one piece and imagine it as *being whole* and therefore separate and unimportant to them. Can you imagine how different the world would be if peoples and nations first looked to those things that join instead of those things that separate?

When something new manifests quickly we tend to think of it as a creation, when it happens over a long period of time we often call it evolution; yet they are born of the same need to unfold. And what is the real difference other than our own perspective of time? When the Wright brothers invented the airplane it took as its basis all that was known about flying and gliding; it stood on the shoulders of all the successes and failures that had come before. In that way each flying machine was connected to the one before and ultimately to the wing of a bird from which Leonardo Da Vinci made his famous drawings. Their vision was brought into reality, but as an aircraft it was only able to fly at very low altitudes for very short distances. In the course of flying the creation, its limitations were discovered and corrected until we have the airplanes we see today. Yet who would say that the process is complete? As we enlarge our vision of flying machines, we will continue to create. We need to think of creation as an ongoing process where we can look back and discover its roots but understand that it is not an end product. We are not able to see an end of creation, only a direction.

The Earth as a mass of material has seen its own lengthy and elaborate evolution. Billions of years ago in a slow-motion shopping spree it gathered the gas and dust about itself in the name of gravity and began to temper, to cook, and to distill its contents. And the three phases of solid, liquid and gas began to form the ancient Earth. But the Earth is never still; change is the common heartbeat of the Universe and the ongoing heat from the sun—together with the residual heat of formation stirred the inner Earth causing oceanic and crustal plates to move. Mountains raised and rivers began to flow and the face of the land had a new paintbrush and so it continues today, an unfinished creation, a work of continual unfolding.

DANCING BETWEEN UNITY AND SEPARATION

In the cycle of life we see over and over again, children who can hardly wait to leave home, only to return at some point, older and wiser with stories to tell of life and their experiences. Joseph Campbell recognized the importance of the quest in all of us that moves us from community where there is connectedness and safety...out into the world. Here great risks are taken as life experiences move the individual beyond familiarity to a place where growth occurs. A great truth is learned at some cost and the person returns to the community to share this knowledge and insight. This is the fabled "hero's" journey in our mythology and variations of this story steep their flavors through our literature and our poetry.

Eventually we all begin to recognize that during our lives we move many times from a sense of unity to a sense of separation and back again. In our individual journey the separation is a state of consciousness that can allow us to focus, concentrate and experience as an individual being. And we realize those times can offer important growth. It is during this experience that we ultimately learn to create balance in our lives not by eliminating opposites but by recognizing them and understanding our connections to them.

It is the feeling of separation, this separation consciousness that also allows us to act for our own short-term selfishness at the expense of others. It causes us to ignore our connections to each other and the Earth and to forget that our futures are entwined. Our sense of separation can lead to a remarkable lack of awareness.

THE BEAVER AND
MY LESSON IN AWARENESS

When I was in college and new in the study of geology I participated in an interesting field experiment. The intent was to show that as stones are swept downstream they are rounded and changed, largely from banging against each other, and those changes in their shape determine how easily a stream can continue to move them. It's a simple argument. Rough rock and stones fresh from being dislodged from some hill or mountain side have numerous sharp edges, and they don't move easily under low to moderate force. Pushed downhill by gravity and storm runoff, a stone eventually finds its way to the bottom of a stream bed where it sits.

Flowing water and the impact of other stones will act to round the edges a bit but only very slowly. It may be there for many years until a storm occurs and the fierceness and energy of the water is sufficient to move the stone. When the edges are finally rounded the stone moves downstream much easier and continues to bounce off other stones that round it even more so it's a supporting process. It is really much like life: we are shaped and rounded by contact and sometimes conflict with those around us until our rough edges are smoothed and we can move along more easily with others. And just as in nature we often experience our biggest movements during the storms of our lives when the energy of change is greatest.

We had collected a number of stones of different specific shapes then recorded their measurements and placed them in a small stream near the campus. Each student painted their stones and then numbered them for identification. The idea was to go back to the stream and find the stones and see which shapes had moved the furthest so we laid them out evenly across the width of the stream in an imaginary starting gate.

I returned to the stream about a month later in the late spring. It was a beautiful warm day and the shade of the overhanging trees was welcomed. An occasional faint breeze cooled my back but otherwise there was stillness and complete quiet except for the sound of the stream and ever-present birds and occasional insects. I found the place of my original "starting line" and began wading downstream in water to my knees looking only directly in front of me through the clear gently moving water at the sandy bottom searching for my stones.

The area along both banks was deeply wooded; a variety of trees including water oaks threw limbs out over the stream and vines and other low vegetation provided a hiding place for a host of animals and birds. In that part of the country snakes were very abundant and most of them were poisonous. Water moccasins were particularly aggressive snakes and in my youth I had seen many, stumbled over more than a few and was actually lucky I had managed to grow up without feeling their bite. Even though they were in the back of my mind I drifted gradually into a very focused state, aware of little more than the search for painted stones on an enjoyable afternoon.

In the middle of my quiet observations I was suddenly stunned by an explosive noise only a few feet in front of me that shattered the stillness like a bolt of lightning on a clear day. It was a beaver that had seen me coming and he had no doubt preferred to stay hidden. But my slow back and forth approach had taken me one step too close and he let me know it by pounding his tail upon the water. I had no idea a beaver could make a sound that loud; I am certain my heart stopped cold and I honestly wondered if it would start again. I froze in one locked-up lump of muscle and adrenaline. Beavers can deliver a nasty bite but seldom choose to do so unless cornered. This one moved away in a moderate haste looking back over its shoulder with what I felt was a look of chastisement. I stood there more shaken than I had been in many years, totally surprised by this encounter and the suddenness with which this beaver had brought himself into my awareness.

The beaver and I had been very close physically, and getting closer, so that separation was hardly there at all in the material sense. The beaver was in the same time and the same space as I was but I was not aware of it at all. And so to me it did not exist and was not important, yet it made itself important to me very quickly and I realized that I was not alone and that my actions had impacts to the beaver and its actions to me. The point is that until something enters our awareness it simply does not exist for us; it could be a million miles away or two feet, it doesn't matter.

When we set aside our lens of separation and begin to look for connections we see the Earth in an entirely different way, an extraordinarily intricate and interdependent system and a ready source from which we can learn a great deal.

CHAPTER 4

THE POWER OF PERCEPTIONS - WHOSE REALITY IS IT ANYWAY?

We are moving steadily and with the shock of awakening into a new realm of deeper questions that center on the nature of who we are. It is a realm where the people and places are the same but our perceptions are vastly different.

There is life out there, abundant life, through any window on every street and extending out to every forest, crawling, creeping and squirming along on planet Earth as it spins through space and moves through time. Up there, out there in the Universe, spread through the heavens of almost infinite magnitudes, we are reminded of the perspective of the long term where there are worlds to explore among the contrast of bright lights in dark skies. Down here back on Earth, there are movies and ball games to go to, ice cream to eat, boys and girls to fall in love, pleasures of the flesh and of the mind. But with such an unlimited close-up carnival, why is it that the older we get the less we seem to actually explore? We tend to wait—stuck in the lives of repetition we have created, caring only for our now-familiar sights and sounds. Is it simply that the young have more energy or have we lost the ability to perceive new things and experience the wonder of life?

For most of us something starts getting lost very early, perhaps between 4th and 5th grades, somewhere between ice cream and girls. Maybe we became too focused on the task at hand and later on too cynical of the world around us, the buying and the selling, work hours, tax time, kids, responsibilities, and then all too quickly it begins to dwindle until it is finally lost for good, leaving us to wonder how life passed us by.

Wouldn't it be nice to see the world again with fresh eyes and feel the wonder of being alive? In the beginning we were all in the main flow of life's great river, but most of us inevitably drift to the quiet and predictable water along the edge. Ever notice the eddy currents along the banks of a river? They go round and round, but never seem to move significantly with or against the river. Sometimes our lives can feel that way, stuck in useless motion that takes us nowhere.

But there are exceptions aren't there? Why do some rare people retain such a life spark, like the fictitious Peter Pan, never growing up and always seeing possibilities? Each one of us has this wonderful little miracle machine we call the brain that helps us survive. It keeps us from being eaten by the dinosaurs of our day, and it can also help us transcend and raise our consciousness. Yet as it processes information to make sense of the world around us it quickly turns our moments of newness and wonder into less exciting experiences of familiarity. According to an online source, The Brain Wizard (www.TheBrainWizard.com), the brain receives about 400 billion bits of information each second— that's an awful lot of data, much more than we can actually bring into our awareness. So the brain filters and categorizes most of the information we receive sending only about 2,000 bits per second into our conscious awareness. How does this work and who's responsible for setting the filters?

The Child Trauma Academy (www.ChildTraumaAcademy.com) points out that, "throughout life the brain is making memories that correspond to various sights, sounds, smells, tastes, and movements. It creates templates of experience against which all future experience is matched." This means that the brain takes each new bit of data and compares it to past experiences trying to assign familiarity to it. In essence the brain matches each event against the library of our experiences and sends the results into our awareness before it moves on to something else.

We are seldom aware of the details of a whole picture... our awareness never "sees" them. Instead we see just enough for the brain to recognize and match a pattern and then we see the *pattern*. It's a lot like the auto complete function on your computer. If it has four legs, fur and purrs, your brain sees the pattern of a cat. And here's the important thing--when your brain identifies the pattern as a cat, it also makes instant associations with past experiences of cats--were they friendly, were they fun, am I going to enjoy this? You see the brain transmits its perception of a cat including *its judgments*. And it's all those judgments that can lead to problems.

Early peoples surviving in small, closely-woven communities recognized that personal conflicts could be catastrophic to the community as a whole. They recognized that each individual had their own perception of events and that a shift in perceptions had to occur to open the door to forgiveness and harmony. To provide a context for this process some early peoples developed what are termed harmonious methods of conflict resolution. There is an interesting saying that comes from the ancient Hawaiian method of harmonious conflict resolution called *Ho'oponopono* that says, *"Changing perceptions allows people to see different things."* They recognized that there are some things you just can't see at all until your perceptions change. In this simple but powerful statement, they have waded deep into the murky waters of our perceptions.

PERCEPTIONS ARE THE BUILDING BLOCKS TO UNDERSTANDING —OR TO MISUNDERSTANDING.

Our perceptions are the filters through which we see the world. Beginning at birth they are formed, nurtured and tuned to be an inescapable bit of our baggage and they follow us everywhere. Unseen and often unrecognized they are the paint we use to color our world. Open, they can be an avenue to exploration, but closed and rigid they can be a stifling force that reduces our life experience and corners us in fear. Like glasses we cannot easily remove, they change our view and direct our feelings and responses. Wouldn't everyone love to see the world with fresh eyes? Yet we are caught in the web of our own experiences, held by each sticky judgment. Our judgments always come back to us in time, sometimes in ways we don't recognize, little incognito emissaries of the past.

Every day is filled with events of one kind or another, happenings of all sorts, some pleasant, some unpleasant. Our perceptions, the way we see these events have a lot to do with whether we find them exhilarating and magical or fearful and restricted.

With persistence we can recognize and change our old limiting perceptions. Wisdom changes perceptions and that is the beginning of a road more open to adventure and a new understanding of our relationship to Earth and to each other.

Occasionally I get the chance to do little outdoor workshops and I sometimes ask the group to look at something, a tree perhaps, and then tell me what they see, one answer from each person. I get many answers—some expected but sometimes a few surprises, too. Often people will answer in outline form, "I see a tree with many branches, reaching very tall, with colored leaves." Or I get answers that center around the apparent health of the tree, how green or how brown. Most everyone grasps the overall beauty and power when paused to look.

At the end of the answers I often ask, "When you look at the tree do you see the insects that live on it, some helpful, some hurtful, some of them on and some of them burrowed beneath the bark? Do you see the bird that feeds on those insects and uses the tree for its nest? Do you notice that the vegetation below is often a little different, of a kind that flourishes in the shade of such a tree? Do you notice that the tree is spaced just-so from its neighbors in the soil? If you look with your mind do you see the deep and powerful roots below that allow the tree to reach so high? Do you see the tree breathing, providing oxygen to our atmosphere? Rooted in its source it is a living organism connected to others, can you see it now?" For a tree is never just a tree standing alone, it is a carnival of life, a pulsating system of inter-dependence, linked in history to the nature of our atmosphere and therefore to our own existence.

Of course people ask whether I can really see all these things in the tree and I smile, for it is the essential question. In this simple exercise we are doing something we talked about earlier, seeing a piece…and imagining the whole. And that little example of what scientists often do is exactly what we do in much of our lives. How open we are to this and how well we do it impacts our life experience. For while the surface that we see may be simple; what we perceive or imagine beneath is often far more complex.

This is particularly true with planet Earth, rich and variant, where you find most anything you seek. Have you ever heard the expression, "Reality is not as important as your perception of it?" This resonates with many of us because the way we see and understand things internally largely determines what we see "out there" in the external world.

The history of scientific exploration is richly colored in examples showing that the way we see things—and therefore the way we search for things— determines a lot of what we find. Science requires testing and historically we tested our ideas the best way we knew how; but after all, they were *our* ideas and we all tend to look for proof that our ideas are correct, not incorrect. And our ideas stem directly from our perceptions, our frame of reference—the way we think.

We spent centuries believing and attempting to prove that the Earth was the center of the Universe, largely because it reflected a view of ourselves as the center of the Universe. We also tried to prove that the Earth was flat and that it was created in 6 days, and that the continents never moved, and on and on. In addition to our perceptions coloring our view of the Earth, they also colored our relationships with other living things. The story of how we viewed dinosaurs is an excellent example. Our early view of dinosaurs could be summed in two words: slow and ugly. We pictured them as cold-blooded and way too big with way too little brain to do much but eat and sleep. It was a picture we created beginning with a few bones and tracks, mated with our perceptions of the past. But that picture has changed a great deal. Now we recognize them as swift, certainly cunning, often partially warm-blooded and in some cases nurturing. They didn't change at all but our perceptions certainly did.

When most of us look up into a spectacular moonless night sky filled with stars, we immediately seek to assign some sense of order. We begin to pick out particular stars and planets, and many people identify constellations. Often the night sky in the Colorado high country is spectacular, particularly during the time of a new moon. And on more than one occasion I remember hearing friends say automatically, "There to the south is Leo and Libra, and over there is Taurus." Frankly I have enough trouble finding Orion, yet I look and begin to pull out of the night sky the shapes of our Zodiac. But let's think about this for a minute. These shapes go back to Greek and Roman times and in identifying and naming constellations they were pulling from their imagination the shapes of familiar patterns and imposing them upon a night sky.

WE HAVE LAIN UPON NATURE IMPRINTS OF OUR OWN PERCEPTIONS.

Our view in the western world of the night sky is colored by the perceptions of a people and an empire long since gone. What do you suppose Mayans from South America see when they look up at the night sky—Leo, Libra and Taurus? Of course not, they see the same stars but impose a totally different set of perceptions upon them. So how do we walk a path of exploration without falling into the box of our own limited vision?

In the last 60 years we have seen enormous advances in science and technology. Each one was led by a change in the way we look at things and in each case, when our vision could accept it, things fell into place rather easily. It was in those last 60 years that we gathered enough data from many sciences to support the notion that our great continents are not rooted in one place as they appear but move around with dramatic consequences. Plate Tectonics models embody data from chemistry, biology, palynology, geochronology, climatology, oceanography, geophysics, and paleontology. Each field supplies information and leads to an understanding of the picture as a whole. Plate Tectonics allows us to place my own state of Colorado and the Rocky Mountains in some global context and breathe a breath of wonder at the magic of their creation.

Most people are now familiar with James Lovelock's hypothesis within which the Earth itself could be viewed as one whole interconnected living system named Gaia after the Grecian Earth Goddess Ge or Gaia. It focuses on the idea of a Mother Earth as the sum of both its living and non-living parts. The Earth is not viewed as a rock with life on it, but rather as a living system. Lovelock used the California Redwood tree as a wonderful analogy. These giant trees can grow as high as 300 feet and weigh as much as 2,000 tons. Some of them are more than 3,000 years old. In Redwood trees 97% of their standing tissues are dead with only a small rim of living cells along the outer edge of the wood.

Yet none would doubt that a Redwood tree is a living system and we wouldn't just call the outer layer of the tree alive and the rest of it dead wood. The same could be said for the Earth. The living and the non-living parts are connected and we can view the whole Earth as alive and acting as a single system.

In this model there is a linking of the Earth's biosphere, atmosphere, oceans, and soil making a system that operates within a balance. We can no longer think of separate components or parts of the Earth as distinct and independent. Everything that happens on the planet—has an effect on the planet. When we burn the rain forest the climate heats up in response; new wind and ocean currents form, all in an effort to restore the balance.

If the Earth is self-regulating, then it will adjust to the impacts of man. But any adjustment will have consequences, so what is it going to do?

The term Gaia allows us to put a name to something many already feel to one degree or another. We find in it a way to take another step toward seeing the earth with a sense of community. And that allows us to *commune* with it, to be nourished by it. Through a change in our perceptions we see much more than cold hard stone.

For a time Newtonian Physics was the best tool to describe and explain the world around us. Yet it was more than a science; it represented a way of understanding the Universe that reflected our own perceptions. At its core was the concept of separation of matter and forces, actions and reactions, and it sought to separate the observer from what was going on all around. As we learned more about the Universe and began to ask new questions, we found that it just didn't explain enough. We needed a larger vision that accommodated the things we see around us with our changing perceptions.

Quantum physics allows us to see the Universe in a different way. It directly incorporates the observer as part of the process, pulling one reality out of infinite possibilities by the simple impact of his/her own observation. The more you read of quantum physics the more it can sound like co-creation, the observer creating a reality out of a Universe of possibilities. That sounds very spiritual. But perhaps that shouldn't surprise us. We as a species are seeking truth and if there is one truth, then ultimately all "wisdom traditions" should find the same truth. In this context science is simply our latest wisdom tradition. Regardless of our path to truth, we get better ideas when our perceptions change and we raise our consciousness.

Our test will always be whether we are attached to the truth or to our own vision of it.

Let me give you an example of differing perceptions. Two friends who are out on a vacation come to the Grand Canyon; they both walk out to one of the lookouts on the South rim. As they approach the edge they both see the enormity, the depth of the canyon, and each feels the same tingling sense of vertigo as they look over the edge to the steep and colorful depths below. At that very moment they begin to travel down different paths of experience. One person feels a sense of fear, his mind calls up momentary visions of falling over the edge to certain death; he grips the rail a little tighter in a subconscious desire to control, to hold fast. Soon his fingers begin to turn white and he breathes only in short shallow breaths. The other person sees the same thing but his mind is filled with the thought of how wonderful it would be to hang glide from this very spot; he imagines that he could stay up for hours floating like the hawks he sees, and he leans over the rail to savor the view.

Both observed the same view, one recognized fear; the other, possibilities. Which visitor would you be? The experiences were different because their perceptions were different. This applies to much more than simple fear. Let me give you another example of different perceptions in the following story:

~ ~ ~

MR. FROG AND FRUIT FLY

Once on a dark and moonless night there was a great croaking going on in a small pond nestled beside a forest. Leaves on the limbs of great trees hung limp in a warm humid summer night air, silent observers cloaked and partially curled in the night. Stillness was the rule this late summer night, but the edge of the water was very much alive, and the air above it. Lightning bugs lit the way for a host of insects wandering in some undiscovered primal journey. A great and green frog of many years lay partially covered in the soft, warm mud. The search for his meal each evening was done with patience and a keen eye and ear. He knew this spot well; it had been his home and his comfort for many years. Though the pond offered plentiful food, he was lonely, frogs do not often find companionship in their own species, they don't team up and go search for dinner. As his loneliness grew he looked for company. His keen eyes noticed a fruit fly landing lightly on a dried twig in the near grass. Mr. Frog did not turn a moment, for though the fruit fly was easy prey they were small and left a bad flavor.

Soon it would leave, why bother to notice its comings and goings? But Fruit Fly did not leave; instead it settled in on the twig as if it had found a new home. Mr. Frog wrinkled his great frog face just a bit in disdain and rolled his eyes, fruit flies—a stupid nuisance. Mr. Frog knew Fruit Fly only as a nuisance; he had no idea that these particular fruit flies live only three days, that all their life experiences are compressed into a terribly small time. All that can be learned, felt and passed on occurs in only three days. In this short time they miss many things, but find others. Of course, Fruit Fly did not know all of that either, he felt that his time of life was just enough, though occasionally he had heard others wishing to live a few minutes more, such was the common wish of all life. Things always left undone, their importance heavily weighted in the perspective of last moments.

Time passed on the pond, the frog waited and ate as he chose and the night just got deeper. Occasionally he would hear the buzz of Fruit Fly as he moved a short distance and then returned. He seemed totally oblivious to this great frog that could make Fruit Fly vanish with a short flick of his sticky and elegant tongue. Eventually Mr. Frog's loneliness overcame him and he said to the Fruit Fly, "Dark night this, you should have seen the full moon of two weeks ago, it was magnificent, large and heavy, overflowing light to all with eyes to see." Fruit Fly thought a moment and said with great seriousness, "What is a moon?"

Mr. Frog was stunned and thought Fruit Fly stupid. He pursed together his great mouth and croaked in frustration. But he remained lonely and minutes later he said to Fruit Fly, "I suppose you have no idea what it is like when the seasons change and the great snow comes. I myself burrow deep into the mud for warmth against the cold" he said with more than a touch of pride. Fruit Fly paused another moment and said, "What is snow?"

This was more than Mr. Frog could bear; after all it was he who had made the great effort to speak, and to this stupid insect, merely lunch in undesirable clothes. He wrinkled up every part of his face in displeasure and turned his great head to tell the Fruit Fly just what he thought of such a stupid animal—only to find that Fruit Fly had just died of old age.

~ ~ ~

This story is an example of the meeting of extremely different perspectives with a result that occurs too often in the world. They lived in close proximity but could find no common ground; they couldn't get past the things that separated them, in this case a vast difference in their experience of time and their perceptions of the world. Yet both lived and both will die; the elements that make them up are essentially the same, and both experience the world in the form that it comes to them.

How much have our perceptions of the world and our sense of what is important changed over what is really just a few short years? Would our conflicts and social differences seem so important if we had a much longer view of things and had lived as individuals through the last 4 billion years instead of perhaps the last 50; if we had seen the rise and fall of mountains, the cooling of the Earth, the absolute connectedness of all things? What have we missed in our short lives; are there metaphorical moons out there that we have not seen, snow we have not felt just because we live for so short a time?

On a practical level, do we really want to go through our entire lives continually bumping up against walls that we create? Unfortunately most of our history as a species right up until about yesterday has been one of limitations constructed out of our own fear and poor vision with the recurring tendency to seek out data and examples that support our existing biases. Changing our perceptions opens the door to a new view of Earth and a new relationship.

Great questions hang before us as a species and it is our vision that will lead us to answers. They are the same questions of the ages outlined in the new light of our unfolding perceptions. To be the people that we can be, to experience and express ourselves fully in this journey we must examine our perceptions deeply, unpeel each layer and discard those that limit and do not serve us.

So we must ask ourselves again every day—when you look at the tree, what do you see?

CHAPTER 5

IF THERE IS A PHYSICAL EVOLUTION, IS THERE A SPIRITUAL ONE, TOO?

A singular journey down a common path...it is the story of the soul's journey of Spiritual Evolution.

It was out of Africa that our ancestors came, moving tentatively away from the forests that had been their watchful home; out they moved along the edges of the savannah beneath the heat, exposed to the sun and wind, and to predators. They moved two-by-two and family-by-family, often in the footsteps of the one before, fearfully at first but blended at the edges with curiosity. Although it may have been a journey of opportunity and sometimes exploration, it was also a journey of necessity. Climates were in transition, not all in one season but sporadically over many seasons and even generations. The deep and abiding green forests that provided shelter for them and their distant ancestors were in slow retreat, each season of cold a little longer, a little more severe.

Responding to change, the herds along the savanna moved first and then the people following their food source. We may even imagine that each day they woke early and watched for the morning sun to light their path. We do not know that this is true, but it comforts our imagination; for in those tenuous times they were as much prey as predator, living precariously near the edge in nature's balance.

We know much about their physical world and we find evidence of their journeys. They were often scavengers—opportunists who may have gathered fruit when it was in season but with yet no ability to plant and prepare a harvest. They had inadequate protection for extremes in temperatures, no way to store meats for times of want, no good ability to carry water any great distance and only a few tools for work or defense. They were subject to disease, pestilence, drought and famine. Within this feast or famine environment they clung to an often harsh existence that showed little forgiveness for mistakes.

Along the path of their exodus there were no roads, no bridges, and no countries, only the hills, streams, and trails of the herd and other animals; and nowhere in the lands before them did they find others quite like themselves. That was about 50,000 years ago and within a remarkably short time they populated much of the coastal areas of the Earth. Scientists search for the signs of their migration and still puzzle at its rapidity. But it was humankind on the move and we know that energy moves to purpose. They brought their tools, their children and an evolving spirituality.

We have explored our physical beginnings for hundreds of years, captivated in the wonderment of how we came to be and where we are going. We have picked through the dust and the bones of a thousand ancient campfires and searched countless sections of rock and sediment seeking clues that might answer these questions and perhaps tell us who we are. As creatures, we are by all definitions astonishing. Our complex bodies are able to gather and process tremendous amounts of data, not just at the level of the brain, but much lower, even at the level of the individual cell. The magnitude and complexity of the DNA that provides the roadmap for human construction boggles the mind. If we unwrapped all of the DNA in our bodies it would stretch to the moon 6,000 times. It is truly a most astounding accomplishment of nature. The ultimate result of our chemical and physical machinery is that we are able to control our internal temperature regardless of the weather, to gather information from a distance through sight, sounds, and smells, to set and maintain thousands of chemical reactions in our bodies each moment, all without any need for it to enter our awareness until something goes wrong.

THE EARTH AFFECTED BOTH OUR PHYSICAL AND SPIRITUAL EVOLUTION

But life's long and miraculous journey of success was not without its own failures, not without distress. Over and over again new life appeared in a moment of geologic time and whispered its presence as a reflection of the Earth like a soft sigh at midnight, hardly recognizable; then gone, a memory. So went each attempt until a foothold was made, procreation achieved and a lineage formed. And as time rolled on through seemingly endless days and nights, the lives and deaths of this great experiment eventually populated almost every available space like sand on the beach. Each surviving species found a niche, however temporary and lived in a window of time sustained by the climate and the other conditions of the day. Fragile, living within nature's balance and abundance, each form of life aligned itself to the cycles and rhythms of Earth.

Yet life was so very persistent. Within the last 540 million years or so, we can recognize at least four major extinctions that decimated life on Earth. After each period of extinction when millions of creatures died and their remains piled high along the streams, coasts and ocean bottoms, we saw a bloom of new life and an accelerated variety of new forms that filled the available and recently vacated space. When impacts from space objects, periods of increased volcanism or other factors induced climatic change and caused extinctions, the remaining creatures changed quickly to fit the new environment. Eventually the Earth provided a playpen full of plants and animals and it may seem that it left us to work out the relationships...the ways we interact.

Humankind has been around about 150,000 years, give or take. Beginnings are always murky and it's sometimes hard to tell what was fully human. Yet at some point hominids began displaying traits that we would all agree are human. Later, in the cold days of the last ice age early peoples in Europe took shelter in natural caves where they huddled for safety against the weather and predators. And here we find some of the first evidence of burial in an organized fashion. It may seem a small thing easily overlooked, but the implications are profound. Many animals grieve for their lost comrades; elephants grieve deeply for their fallen and are reluctant to leave the bodies for many days, but they do not bury them. These burials may have been some of the first times that early humans did not just grieve and then move on. Perhaps the burials were simply a means to remove a food source and hide their location from predators; but perhaps something more powerful was taking place—a seeping in of consciousness around the campfire.

From our own-recorded history we know that fear is the ready partner of ignorance, and ancient people could have understood their physical world only poorly. Perhaps they spent a great deal of their time anxious and afraid. Fear resides at the boundaries of familiarity, yet we are drawn beyond our boundaries in our exploration and growth. Early humans had a very full day in the heart of nature. They spent their time acquiring food without becoming food and in the in-between times they thought and they wondered.

The northern climates with their long dark nights and bitter cold winters confronted our ancestors with an often stark and hostile environment. The ability to utilize fire for warmth and protection liberated early peoples and provided more free time. And in the darkness of the long nights came the time to imagine and create. Nature provided a quiet and fertile time of the mind in the waiting, in the passing of the seasons, in the cycles of Earth.

In learning about the Earth, how it works and its principles of creation, we also learn about ourselves. Those first steps that humankind took in walking to the edge of their fear and beginning to understand and moderate their surroundings were actually the first steps in an ultimate journey of self-discovery.

In exploring the past we have unearthed the rise of our species noting the achievements to our collective community and our civilization along the way. Throughout our development the Earth was a presiding force in the important factors that allowed our species to progress.

It is our nature to look back along the path of our development and seek that fork in the evolutionary road that led to us, to question what piece of luck or logic caused us to be what and who we are. Our environment changed and changes in the Earth required a responding change in the species. Each adaptation of our species moved us closer to who we are today and in this way it was the Earth that took our hand and guided our development. The migration out of Africa may have been spurred on by a climate change that began millions of years before, when the ancient and abundant forests began a slow and steady retreat. Was this the reason our human ancestors left the safety at the edge of the forests and began a long journey that populated the earth?

Since the time of our beginnings we have grown stronger, taller, faster, and smarter. And yet the last time I looked none would say that the journey is complete, that evolution's pinnacle is attained or that humans have reached a place where there will be no further physical change—no getting taller, shorter, faster, or smarter. We seem to recognize in our core that life is indeed a journey of unfoldment. Would we not expect our consciousness, our view of spirituality to continue to evolve as well?

Our own biology carries the history of our species as a genetic memory. Sometimes we feel like we live in an envelope of intuitions that may be memories of distant times when our lives pulsed in harmony with the seasons and the phases of the moon. Have you ever noticed in the fall that many of us tend to put on a few pounds and want to read or sit by the TV? It's a time when our physical activities are often reduced. And it's a time when our bodies undergo subtle chemical changes; our cholesterol goes up about ten points without any change in diet. And we tend to sleep more. Is this a reminder of the time when our own cycles more closely followed the Earth and we, like the bear, needed to hibernate or at least use less energy in the winter when food was scarce? Are we gently responding to the distant sounds of biologic echoes? Clearly our stay on Earth has guided our physical development and in a similar way our Earth experience has impacted our changing views of spirituality.

EXPRESSIONS OF OUR EMERGING SPIRITUALITY

The migration out of Africa was a journey of the species, a piece of the story of our physical beginnings taken from science. And science reserves the right to change its story when new information is discovered. Among indigenous peoples there remain other stories rich and beautiful passed down from generation to generation that describe how humankind was created and populated the world. And there is another journey, the one within for greater understanding, the one we must all take. Some approach this opportunity for self-discovery with excitement and with focus, most often however it comes over time, and slowly...often reluctantly. And though we try desperately to share this journey and take others along, it is ultimately a singular and internal experience.

When I was growing up religion, which represented all of my spirituality, was presented to me as fully formed, complete, perfect and with no place to go. Yet as our perceptions of the world around us change, so too our views of spirituality will inevitably change and we can only wonder where that journey will take us.

Deeply rooted in our collective psyche, our view of Earth has always been linked to our view of God. Early peoples struggled to understand the world they lived in and their view of nature defined their view of God. And what a frightening view it must have been. Violent storms of all kinds with incomprehensible lightning, booming thunder and sometimes hail—probably instilled terror in the hearts of early peoples, often surviving in poor shelter. Survival had to be wrenched away from nature and at times they were caught in unexpected floods, or searing heat and wind in what must have seemed like arbitrary actions of an angry God. Hungry and perhaps unpredictable animals added to their fear. It is no wonder that their first perceptions saw nature as a dangerous place and their Gods as powerful, puzzling and angry. It was a view painted richly by their fears and uncertainty, and a sense of vulnerability. But early peoples also began to recognize the miraculous transformations that Earth displayed each year: the cycles of the seasons, the transforming of seeds into fruit, the coming of the rains and the filling of dry streams and lakes. All of these things began to give a sense of dependability in an uncertain world.

There are many religions or wisdom traditions that have grown from the peoples of the Earth. At first glance we might believe they could hardly be more different. Some are more ancient, some are more recent, but at the heart of these wisdom traditions lie simple universal truths. And there is another wisdom tradition that we call science, for all are seeking an ultimate Truth.

Wisdom traditions of indigenous peoples speak of a link and a connectedness with the Earth that goes beyond a casual association. It is a fundamental and treasured relationship. Buddhism focuses on self-awakening but it too recognizes the connected nature of all things and a reverence of the Earth in its "Touch the Earth" practice where individuals kneel and touch the Earth in meditation. It was said that Siddhartha had moments of doubt as he neared ascension and that he practiced Touch the Earth and ascended the next day. Buddhism recognizes that none stands alone; all are connected to the past and the future.

The wisdom traditions of early peoples speak of an ancient and connected past. There is a relationship of Earth as Mother, sky as Father and a vision of life in everything— in the mountains and waters as well as the animals. In this way they have always seen the Earth as alive. Indigenous wisdom seeks to live in balance with an Earth that is older than prayer. Shamans sometimes think of themselves as people who deal with the tears and holes in the net of life and in a larger sense attempt to recognize and restore balance between the material and spiritual worlds.

Native Americans speak beautifully of their deep connection to Earth. In the book, *Touch the Earth: A Self Portrait of Indian Existence,* Chief Standing Bear of the Lakota is quoted as saying "It was good for the skin to touch the earth and the old people liked to remove their moccasins and walk with bare feet on the sacred earth. Their tipis (sic) were built upon the earth and their altars were made of earth. The birds that flew in the air came to rest upon the earth and it was the final abiding place of all things that lived and grew. The soil was soothing, strengthening, cleansing and healing. That is why the old Indian still sits upon the earth instead of propping himself up and away from its life-giving forces."

There are striking commonalities in the spiritual practices of ancient peoples. Since they often developed separately without significant early communication we can sense that it is a common response to Mother Earth from deep within the indigenous soul.

Community is sacred in an environment where everyone was important to the survival of the whole. The community often consists of those living along with spiritual ancestors. Spirits of the ancestors are called for ceremonies and important decisions; honoring the continuity of generations emphasizes a sense of connectedness.

The concept of the sacredness of sound permeates many spiritualities. From Native American tradition we learn that, "Song is the breath of Spirit that consecrates the act of life." The King James Bible says, "In the beginning was the Word and the Word was God." "The word" can be thought of as breath. From the Hindu we learn that out of the vibration of Nada (original sound ether) comes the Universe, while Mayan tradition teaches that the other world sings us into being and that we are its song.

Two Creation Stories

Just for fun let's compare two different stories of how things are created from very different cultures and see if there are any similarities. There is a particularly beautiful understanding from the Mayan culture that is simple and profound. Metaphors have always been the vehicle of the heart, the most intimate way we grasp the awe-inspiring and pull it close. Mayans imagined two worlds in balance, a spiritual world and a material world. They envisioned this like a tree where the visible portion of the trunk and all the branches above ground are the material world and the invisible roots holding it up and nourishing it represent the spiritual world. In order to grow "tall" in the material world we must always deepen our spiritual roots or the tree falls in on itself. A Mayan story says that as sound passes through the sieve between this world and the other, it takes the shape of birds, grass, and trees, etc. (*The Sun* Interview with Martin Prechtel April 2001, "Saving the Indigenous Soul" by Derric Jenson). In this simple creation story there is a sense of Spirit in sound, and sound translating into form.

And here is another understanding from more modern times. In his book, *The Fabric of the Cosmos,* Brian Green discusses String Theory and how it seeks to answer the deepest questions of the cosmos; questions of how all the largest and greatest things came to be. However it starts by looking at the building blocks, the very small things—smaller than atoms and subatomic particles—so small in fact that at some point the theory predicts there can be no smaller particles, only "filaments" of vibrating energy.

And the theory proposes that the nature and frequency of that vibrating energy determines what kind of particle of matter is formed from it. Very simply put, in the view of String Theory, there is a vibration of energy that is translated into subatomic particles which then form atoms. These atoms go on to form molecules which ultimately form the same birds, grass, and trees we remember from the Mayan creation myth.

They are both stories of creation—one from the heart and the other from the head—both seeking truth and providing explanations for the wonder around us. The core of the two beliefs resides in the vibrating movement between matter and energy, and metaphorically between Spirit and form.

Spirituality is an experience of mystery and it is a basic element of the individual journey. As we explore spirituality we find ourselves looking again at Earth. Nature does not deal in morality, but rather in the underlying practical functionalities.

Ultimately nature does not judge right and wrong, only what works and what does not. Yet within the principles of what works we find reflected values that can be applied in our own lives.

In our beginnings we identified ourselves with our immediate needs: food, water and shelter yet there are emotional and spiritual needs as well. An unfolding Earth energized by the sun provided the necessary foundation for the development of our psyches as well as our bodies. Ancient peoples lived in a close emotional connection to the Earth that went well beyond seeing it as shelter and a place to find dinner.

Of the wisdom traditions, Shamanism is the oldest and science is the youngest. But whether we seek truth through ancient ritual or through a scanning electron microscope, we find connection to Earth, physical and emotional. Ancient peoples would say that it is simply a case of the children of the Earth knowing who their Mother is.

It was not surprising that early peoples worshiped aspects of the Earth. The concepts of gratitude, love of beauty and appreciation of grandeur were born in such things as a rich red sunset, the relieving peace that follows the storm, and the deep stillness of newly fallen snow— pure, unmarked and waiting. But planet Earth is not just a "rememberer," holding secrets of our past, but also a "revealer" where we can see the basic principles of creation in action.

Our physical evolution took billions of years but spiritual growth accelerates rapidly once the spark of consciousness becomes flame. And in the last 50,000 years humankind exploded in an exponential growth of consciousness.

In the course of our spiritual evolution we began by searching outwardly in the Earth for clues of our own being and found reverberating answers within. We changed our own perspective through quiet hours of contemplation in the rhythmic centers of Earth where its beauty and grandeur lift to higher consciousness on wings of the soul. Great mystics, prophets, thinkers and leaders have all periodically gone back to nature. And they did not go there just to *get away*—they went to *get* **to**. For in those quiet, reverent places, Earth provided beautiful and endearing imagery that touched the heart as well as the head. And in observing Earth we found inspiration for everything from art to engineering. In our own reflection we intuit what the Lakota knew, that as we distance from the Earth connection, we distance our humanity as well.

There is a yearning in each of us for this connection... perhaps it is buried deep within our indigenous soul, long forgotten but not lost. The awakening of our species came in part through the exploration and understanding of this mystery we call Earth. The millions of years of time and the testing of our ancestors pushed our ability to grow, to expand and have resulted in finding ourselves where we are at this moment. We are not at the end and not at the beginning, but somewhere on a never-ending journey of understanding: of ourselves, our planet and our place in the universe.

In the quiet places of Earth—perhaps beneath the grey Spanish moss, in decaying layers, flavorful, steeped in the smells of death and rebirth, or alone in the raw and colored desert, or by the sea, or along a mountain ridge; in these places and many more Spirit speaks its name for all to hear. A changing Earth guides and embraces a blossoming spirituality; it sets the stage for our learning and lays upon us a direction. It provides the stimulus for which we provide the response. Entwined in all of this we see unerring principles of creation that brought the rock in place and developed the life on it not as separate events but as a continual unfolding.

Our ancestors felt the same things we do: fear, wonder, doubt, joy. Yet when they reached down and put their hands in a rich soil they felt a connection to the essence of life itself. And then they proceeded to build a world. For people are above all optimistic, seeing the potential of what can be. We are mixtures of action and feeling, of hope, inner strength and vulnerability. When we deeply touch these facets of our being we are most truly alive.

In our exploration of the Earth we dip into history's cup seeking an ancient balance upon which we can raise a new and greater vision. We do not seek to create the vision; we seek to release it.

CHAPTER 6

WHAT VALUES DOES
THE EARTH SUPPORT?

Straining in the night for glimmers of direction, actions to take, we discover it is not force but understanding that releases. We do not pull the grain from the Earth; we prepare the soil and wait for Her season.

Sometimes gently, sometimes with great power the Earth reveals its values, the underlying principles that sustain. It's not just the bright signs and flashing lights of Earth's more impressive messages that concern us; it also expresses itself in more subtle patterns that can be felt in our deepest core long before they can be described in words.

Many of us have heard the saying, "Align with the design." This simple yet powerful bit of deep wisdom conveys the heart of the message. For nature does not judge, does not conspire against us. Rather it acts along straightforward principles that form the design of Mother Earth and they are a design for success in our own lives. In understanding and accepting the rules of the design, we are in tune and more at home with the Earth and with ourselves. We remember that when we look at Earth we are looking at something created in the same factory as we, and by the same tools and principles. When we can grasp this powerful and deep kinship, we see a world of almost infinite variety, yet all manifested by the same underlying principles.

The Earth, keeper of our past and holder of our evolutionary story, contains all opposites within its long experience: violence, suffering, hope, and love. So it is not surprising that if we look deeper, we can see analogies and metaphors for almost every situation, every occurrence in our own lives; some of only passing significance, others of great importance. Sometimes we can see events in the Earth that don't always seem to make sense—that appear arbitrary. Yet over the longer term, we recognize patterns in the fabric of creation that clothe and support our underlying foundations and give us an anchor, a sense of consistency.

What handful of core values does the Earth support that could be useful to us and how will we know them? Life continues to appear in new and ever-changing forms and it can be a sloppy process, exploring all possibilities; yet underlying principles will continue to decide success or failure over time. So it is the principles we must recognize and we find them by looking at Earth, the way it has evolved, in the building of the mountains and the oceans and in the successes and failures of the creatures that inhabit them.

The varieties of life that have lived on the Earth are astounding and we see in their histories ideas that worked wonderfully well and many that didn't, the life forms that were discarded along the side of evolution's road. Buddhist teaching says that when conditions are right things will manifest, and when conditions are not right, things will withdraw. Truly that seems to describe the history of life on a changing planet. The Earth is an organic remembrance of the creation process and it holds within it all the values of its creation. As they are applied in its own development it creates for us a cosmic roadmap for our own lives. So again, in looking closely at our Earth, what do we see?

THE VALUE OF EMBRACING CHANGE

Embracing change is such an incredibly important key to our happiness and well being, yet such a very difficult thing to do. Within the zone of Earth's great balance, change is an ongoing and inevitable process; it is perhaps the one thing we can count on. And not just a little change every once in a while, I mean the significant life-affecting changes that we all experience.

As a species we love to explore and experience new and different things, but there is another part of us that is an extraordinary nest builder. We like for things to be comfortable, safe and predictable, and the older we get, the more we rely on the nests of familiarity we have built and the less open we are to change. On some level we like to know exactly what this evening and tomorrow are likely to bring and we work very hard to create that stability, that predictability in our minds if not in our reality. Subconsciously our brains like to visualize things and we can easily visualize the familiar—we have seen it before.

The Child Trauma Academy (www.ChildTraumaAcademy.com) notes in its on-line lessons about the human brain that, "...the brain is a conservative organ. It does not like to be surprised. All unknown or unfamiliar environmental cues are judged to be threatening until proven otherwise." When we think of going home at the end of the day for instance, the brain not only recognizes that activity, it may envision a familiar and favorite meal with friends and family. The association it makes is one of comfort and that relaxes us and reduces stress. As we learned earlier, the brain is always comparing the information it is receiving against its library of experiences, searching for familiar patterns. Diane Ackerman in an excerpt from her book, *An Alchemy of Mind: The Marvel and Mystery of the Brain,* for the June 15 New York Times, pointed out that, "Pattern pleases us, rewards a mind seduced and yet exhausted by complexity."

Change requires visualizing possibilities and that is far more difficult. The brain has no familiar vision or association. When change means giving up the known for the unknown, it can be very scary and stressful.

Yet we know that nature is constantly changing. Every single day our environment is remolded and reshaped; rivers around the world dump thousands of tons of sediment into the oceans, eroded from the mountains and other high places; they are filling in the lows in an inexorable movement to balance. How can we best prepare for the change we know is inevitable? It sounds inviting to think of embracing change, of telling ourselves that in every time of uncertainty there are opportunities for growth, right up until the moment we actually have to do it. Then our fears take over and we try our very hardest to return to the comfort of predictability, even if it is limiting and unsatisfying.

The Universe will have none of that and despite our strongest resistance and greatest efforts to cling to the present, change rains down upon us. Our challenge is to accept and explore these changes without losing our sense of who we are and our long-term goals. As we begin to grasp the depth of the change, its impacts and its possibilities, we start to form a new comfort zone, and regain our balance.

It is sometimes exhilarating and often painful, but it is certainly a never-ending process. In fact some would say that when we quit changing we start dying. Change is necessary for growth to occur and there is another aspect of change that is also important to our wellbeing—the letting go. We are all aware that snakes shed their skin when they have outgrown it and it is no longer useful. As the nautilus grows it closes off older parts of its shell and creates a new and bigger living space, one more suitable for its growth. When crabs and lobsters outgrow their shells they step out of them, even though they are vulnerable to predators—until they are able to grow a new one that fits better. They have risked their very lives to grow. In these examples of change, letting go was necessary for the growth to occur. They had to discard those things that could not grow along with them and this is a critical factor in embracing change. We must let go of things to create room for growth.

In looking at the long term, people may notice that times of change often come in cycles like the seasons. Whether the changes relate to our relationships, where we live or our work and our life passions, each involves a need for giving up the old. And we often need a quiet time of renewal during which we re-connect to what is important before we can enter a springtime of new growth.

Clearly change is less scary when we have an idea of what we are changing to, not just what we are changing from. And if we recast the idea of change in our minds as a transition, a shift, it does not seem so random and can be less scary. When stress is reduced, we can begin to look for the gifts in the change, for there are virtually always gifts. It is much easier for trapeze artists to let go of the bar that supports them if they can see the bar they are reaching for. It's not always that easy in our lives, but a focus on what we are becoming and a comfort in the certainty and natural order of Earth's changes make the task a lot easier. During times of change it also helps us to strengthen our attachments to things that are timeless.

Our connection to the wonders of nature; to the oceans, the mountains, and the plains forms an anchor of certainty in times of change.

THE VALUE OF SIMPLICITY

Such a rich and elegant word, simplicity; say it softly and it slips gracefully off your tongue. Simplicity is unpretentious and has an implied directness, as if exposing truth to the light of day. The word connotes a sense of ease, less energy to maintain, the uncluttered approach, nothing frivolous and unnecessary, few distractions. We may associate simplicity with an un-gilded and clear purpose and a straightforward action of no deception that brings light into a dark place. What images come to your mind? I can easily imagine being away from the noise and distractions of civilization; a warm summer afternoon spent relaxing as a breeze drifts gently through the trees and the leaves murmur in quiet response. But I would be mistaken in confusing a sense of calm for simplicity. We can find lots of activity and sometimes aggression in nature's expressions of simplicity, as we shall see.

Geologists dig deeply through the past, sifting for clues that help us better understand our world and the history of our species and sometimes they find stories with a common theme that are repeated over and over again. History is full of examples of species that began as simple creatures—that found an environmental niche where they could survive, then grew into that niche and later flourished and became abundant. This sounds like another evolution success story, however the process of change and adaptation did not stop. Some flourishing marine micro-organisms had more food and energy than they needed to survive and in these "best of times" they used that energy to develop shells that were very ornate and often very beautiful. Eventually over thousands or millions of years they became highly specialized and dependent upon the subtleties of their environment. They may have become attached to one particular food source that may have been high in energy and easy to catch; often they came to require a certain temperature, a particular amount of sunlight or moisture. This works wonderfully well as long as things don't change. But we know that nature is married to change and sooner or later either the climate got colder or hotter, wetter or dryer, or their food became scarce, or a new predator came into town. Burdened by the specialized requirements of their lifestyle or having no defense against a new threat, they were unable to adapt to the change and many species simply died out.

Those critters that need the least adapt the best to major change. Good examples of this are mosquitoes and sharks. They seem at first so different but they are connected in their simplicity. Both have been around hundreds of millions of years and both have changed very little. Mosquitoes have a simple design and from an energy point of view are very cheap to "build." They do require sufficient, still, fresh water to lay their eggs and if the water dries up you may lose a generation of a billion or two but it really doesn't matter; they breed quickly and they will be back in overwhelming numbers.

Sharks are certainly much more complex than mosquitoes, but their internal support structure is relatively simple compared to other fishes. After all, they never made the evolutionary jump to developing bones instead of cartilage. Their entire body of hundreds or even thousands of pounds of muscle is able to travel at high speed and turn quickly supported only by cartilage, similar to that we find holding up our own noses. There is nothing more graceful yet more frightening than a large shark in the open water. Both sharks and mosquitoes survived through times of significant change and they are extraordinary examples of being an elegant solution without being "ornate".

There are other examples. Army ants found their niche long ago and have evolved little or none at all since the age of dinosaurs. Army ants do not share the mobility of mosquitoes and sharks and yet they are similar around the world implying a common ancestor. Purposeful, driven by a strong sense of devotion to their community, they have withstood all the climatic changes and stress of an evolving Earth with no significant change. Those life forms that are simple at their core tend to last through change. Likewise, being ornate or overly specialized to any particular environment is dangerous in the long run.

The Earth speaks in simple tones reflected in our own desire for an uncluttered, straightforward and understandable world. We find that life in general and change in particular is handled so much easier when we carry less of a burden, less of those things that are not absolutely necessary to our happiness, when we have fewer commitments, fewer requirements to support. It is a simple but powerful message. We hear so much today about the need to simplify our lives.

It means that we need to decide in our lives what is important, and what is really just ornamentation.

THE VALUE OF FLEXIBILITY

We know and understand that change is going to come, sometimes in a helpful way, but sometimes in a stressful way that tests us to our core. And to add to the difficulty, we tend to become set in our ways, comfortable in our nests; we don't want anything to disrupt our familiar patterns. We all have the ability to set our own course as we go through life yet as we do, we must remember that the ocean is moving too and course corrections will become necessary.

Being flexible is a powerful tool that helps us adapt to changing conditions. As the creatures of nature have discovered, when things keep changing it really pays to be flexible in terms of what and how much you need to eat and drink, or how much sunlight you need, or how much warmth. If you live only on Snow Geese and you really like the taste, what do you do when they all fly south? We all know that when the path is blocked, a river is completely flexible and seeks a new direction. The force that drives it on does not stop twisting and turning until it finds its outlet. Nature reminds us there are many paths that lead to the same outcome, and often new paths carry unexpected rewards. All of these things make us aware of the need for flexibility and cause us to periodically question are we flexible in the way we live our lives? For when the winds of change blow hard, even the greatest of trees must bend or they will break.

THE VALUE OF DEPENDABILITY

Earth's interdependent living systems are often anchored by events that are dependable year after year. This may sound strange after recognizing that change is a pervasive and inevitable part of out lives. But in a macro sense, life systems grow around the dependability that the sun will continue to give its light and warmth, and that the cycle of the seasons will occur on schedule every year. Grass and other seeds begin their growth in the spring long before we can feel a warming and in so doing are already sprouted and ready to make use of the sun's intensifying rays when they finally beam fully down.

It is not difficult to imagine the catastrophe that would occur if seeds used their limited and precious energy to sprout only to find the sun's anticipated warmth long delayed.

Animals that migrate often begin to move well before the seasons change in the same way that many flowers open before the rain comes. And migrating animals usually depend on an abundant and easily attainable food source at the end of their long and tiring journey. Exhausted, they must replenish their systems. In the case of birds, this may be a succulent berry and if the migration is too early so that the berries have not formed, or too late so that the berries are gone, the result can be starvation and death. In cases like these, dependability in nature's signals allows living systems to anticipate and prepare for change.

Snow accumulates at the higher elevations in the Colorado Rockies through the long winter months. The snow contains a tremendous amount of water and it provides much more than just a beautiful recreation opportunity for skiers around the country. When the spring comes and warmth returns to the high country the snow melts slowly and flows down the Colorado River and its tributaries providing needed water for cities and farmlands all the way into California.

It is not just the amount of water that is important but the timing of when it becomes available. Melting snows deliver the water when it is needed for spring and summer irrigation. Should global warming raise the temperature in the Rockies so that the winter moisture appears as more rain and less snow, the impacts would be significant. The rain would flow down the Colorado and its tributaries immediately rather than be stored until the spring melt. Humans as well as a variety of wildlife have become dependent on the timing of this precious moisture.

Without being aware, we make most of our decisions and build our lives while depending on the actions of nature and often the promises of those around us. Dependability provides a sense of certainty in a changing world; it is something we count on. We require dependability in our relationships.

THE VALUE OF CONSERVATION

Simply put, this translates to increasing our efficiency and acting to minimize waste. It means stopping those things that needlessly reduce the amount of our necessary resources, reducing the waste we generate. It means using less water than is available, less food than can be grown, gathered or caught. The Earth is an organic garden but it does not produce the same amount of food each year; rain does not come to each place at the same time or in the same amounts each year, and temperatures also vary, both locally and globally. As we know, there will be times of great abundance and times of drought and famine. You might think of it as another cycle, the Earth's cycle of productivity. Within this cycle we find our own zone of comfort, balancing our needs against the abundance of our resources at the time.

Critters, individuals and even societies that practice efficiency and conservation tend to survive when things change—when water, food, or heat is less abundant. Those living at the limits are the most difficult to sustain and the first to go. The reason is simple—they have no margin for error, no room for change; they are *living at the edge*.

And there are things here to consider in our own lives, too. As individuals we need to question how often we live at the very limits of our time, health and money. We are programmed in our society to consume—to believe that more is better and we are inundated with marketing messages that identify exactly what we should acquire. We are told this will make us healthier and happier. How much extra stress do we put upon ourselves to have that bigger house, the third car? When we live more comfortably inside our resource limits, we find a peace and calmness that allows us to see our choices and our direction more clearly. When we apply it to our planet we call it sustainability which in its simplest terms means evolving and developing our species within the limits of renewable resources. It means ensuring the things we run on, don't run out.

In earlier more simple times most of the resources that a community needed were obtained locally. Crops were grown within walking distance and the food was brought to the center of the village or the tribe and distributed to the community. Water was nearby, as well as the fish that it brought. Hunters needed to go on longer trips to follow game, yet it was still relatively near. In this way the availability of resources actually provided a limit to the population. By comparison, our modern cities pull in resources of food and materials from around the world to satisfy the needs and desires of their inhabitants.

In the past we recognized that the world contained large untapped supplies of energy and raw materials that we could use for many years to come. However, the industrial revolution in Europe and the United States brought tremendous change and advancement and with it a dramatic increase in the consumption of natural resources. Not a problem—we expected that new technologies would take us beyond the need for these critical resources before we ran out of them. This made it a whole lot more comfortable to just plow ahead and not worry too much about declining resources.

Today there is a great disparity among the Earth's peoples, and those emerging populations coming rapidly into their own understandably want many of the things associated with western lifestyles. The problem is that western civilization lives at a level of consumption that could not be maintained for all the people who are already on the Earth. We have imported resources from around the world to raise our standard of living and support our dependencies, and not without conflict.

In our oceans, upwelling currents periodically bring nutrient-rich waters near the surface. Phytoplankton algae in the waters require sunlight and they also require crucial minerals found in the nutrient rich waters. These algae are able to reproduce quite rapidly and they routinely produce enough of themselves to consume virtually all of the available nutrients. This process can result in an algal bloom or what is sometimes called a "red tide" depending on the color. But what happens when the nutrients are gone? Phytoplankton don't have the ability to go search elsewhere for food, they can only drift at the mercy of the currents. And so they die off just as rapidly as they appear; out of fuel they are also out of luck. The loss of billions of phytoplankton can be replaced; however, the lesson for humans is simple and clear.

Even with accumulated information telling us that we must be careful, there remain strong forces pushing us down the same road that some species have taken to living at consumption's edge. Many of those species are now gone— victims of change and it remains to be seen how we will accommodate the warnings. We don't give up our habits easily as we know from the initial strong resistance to reducing the emission of greenhouse gases. The first response was denial and we lived in that stage for many, many years. Political leadership and special interests benefiting from the production and use of hydrocarbons continued to deny the impacts of global warming and even that global warming was in fact taking place. Ultimately we entered a phase of acceptance and with it we have taken a new, more serious look at alternative fuels and conservation.

THE VALUE OF PERSISTENCE

One of the great lessons of the Earth's four and one-half billion year evolution is that things do not usually arrive fully formed. Humankind however, has grown increasingly expectant of instant results and will settle for nothing less.

When I was a child I remember being captivated by magicians. They made things appear and disappear. This became my idea of creation—that it was an instant appearance much like a rabbit out of a hat. But that was illusion, not creation. Oh it's true that some things may be created in an instant—an idea, a smile, a mutation—still, the overriding message is that things take time, often lots of time.

In our lives of instant needs/gratification we go to the grocery store and pick the fully ripened fruits and few ever think much of where it comes from. How many think of the seed, the planting, watering, fertilizing, the painstaking weeding, long days of growing under the sun and the constant care and attention of the farmer? Creation is most often a revealing process of sequential steps where every building block is important and requires persistence to put in place. Each of us lives in the flower of that effort.

I once read a Namaste message that summed up our impatience. It referenced Shirley MacLaine's character in "Postcards from the Edge" as she said to her daughter played by Meryl Streep, "The problem with your generation is that you want instant gratification," while Streep's character replied, "Mother, instant gratification is too slow." When we look at the Earth we wonder, how long did it take to form something as magnificent as the Grand Canyon? Actually we see it only now in this time and in this phase—

it is not fully created yet. The raising up of mountains, the carving of our beautiful valleys, and the growth of our great forests all took enormous amounts of time. The very atmosphere that we breathe each minute was created molecule by molecule over billions of years. The Earth has a wonderful way of providing the necessities for life to thrive, given enough time.

We have a lot of tools at our disposal as we make our own way in this world, creating and fulfilling our dreams. We use our skills, our physical and emotional strength, and our craftiness yet nothing is more important or powerful than simple persistence.

Patience and persistence allow us to look to the long term. There is an ebb and flow in all things and there will be good times and bad. Difficult times can pull us together—they allow us to better appreciate each other and our individual gifts. The time spent in fulfilling a dream allows us to re-evaluate our goals, discard the unimportant and sometimes change direction to something new and better. Do not judge your own life by what you can see in front of you.

Great works take time
great lives do, too.

THE VALUE OF SYMBIOTIC RELATIONSHIPS

Relationships where two different species work together to the benefit of both are termed symbiotic. At first it seems these may be rare—the idea of totally different species with seemingly different lifestyles and different goals developing a relationship that fills a common need—yet there are many around us. One very common relationship many people see every day is that of trees and birds. Trees provide a safe home and a sturdy place to build a nest. Birds often eat insects that damage trees, and they may also eat the nuts and distribute them far away where they may germinate in a new area. These different needs are met in a common solution that benefits both.

Prairie dogs and burrowing owls also have a symbiotic relationship. The owl gets its home readymade from the abandoned burrows of the prairie dogs. The prairie dogs eat away grass near the burrows so that the favorite prey of the owls, insects, are easily seen. Prairie dogs have many predators that rely on them for food, and they are an easy prey if they can be caught wandering from their protective burrows. The owls, like most birds, have exceptional eyesight and as they are looking for insects, they also identify predators. In fact, they both watch for danger and give warning calls that are heeded by the other.

The oceans have their own symbiotic relationships such as the small cleaner fish that pick bits of food and parasites from much bigger fish. To do this, the cleaner fish often swims inside the mouth of the larger fish to get all the areas that need attention. And even though the cleaner fish is working in the very jaws that could crush it, the larger fish refuses to eat it. I have seen this firsthand and it is fascinating to watch: a large, stationary fish with its mouth wide open and a sort of glassy-eyed look, enjoying the attention of a tiny fish that darts in and out of its great mouth. Clearly it enjoys being cleaned more than having an easy meal.

Sea Anemones, stationary relatives of jellyfish, use their poisonous tentacles to capture and devour fish that venture too close; they are not generally particular about which kinds of fish they have for dinner. Yet there is an exception: the small Clown Fish is ignored by the Anemone. The Clown

Fish uses the Anemone for protection, swimming comfortably and undisturbed, deep amid the Anemone's deadly tentacles. Meanwhile the Anemone benefits from an easy lunch of any fish that follows the Clown Fish into its tentacles.

The lesson is pretty simple: do something for somebody—they may help you back and in ways you don't expect. It's not just a motto or a good feeling—the Earth supports it.

How did these interesting and complicated relationships of interdependence begin? There are now many varieties of symbiotic relationships but they didn't start out that way, quite the opposite—they were often parasitic in the beginning. One organism feeds off another like ants feeding on a particular succulent plant. In its defense, the plant begins to secrete poisons that kill insects including ants that crawl upon it. The ants adapt to the plant's defense developing immunity to the poison. The plant then develops a stronger poison and on it goes like some oddly formed arms race. The ants—by their persistent presence and territorial defense—may chase away other insects or other life forms that want to feed on the plant. In some cases the plant eventually accepts the ants and may even develop new odors or flavors to entice the ants who have now become their defender. Thus, a very simple but powerful relationship is born—a symbiotic relationship that benefits both organisms.

Some Brazilian ant-plants have gone much further and provide a large area of accommodation/housing for the ant colony inside a swollen portion of the plant itself. The ants gain a home, while the rootless plant collects the leftover nutrients from the ant's meals and waste.

What we see here is an evolution of the relationship itself from parasitic and independent to symbiotic and interdependent. This movement from individuality to community operates in many levels including humankind. In her book, *EarthDance: Living Systems in Evolution,* Elisabet Sahtouris recognized a repeating pattern in the relationships of individuals that included movements from an initial sense of unity in childhood to individuation to competition; to conflict, negotiation, resolution, and finally cooperation.

Western civilization still largely operates in the phase of competition whether individually or in teams. It is deeply ingrained in our children from an early age and follows them throughout life. Competition, the same valuable principle that helped defend the home, secure the mate and achieve prosperity, does not work as well between communities at large. Competition is sometimes described as an adolescent behavior and it reflects the state of our collective consciousness. As a relatively young species, we stand on the threshold of choice, pulled between the patterns of competition and cooperation.

BALANCE –
THE ENCOMPASSING VALUE

Balance is such a subtle yet powerful word. What does it mean to you? What do we find in a state of balance and why would we want to be there? Balance can imply to many people that nothing is happening, when in fact a great deal is happening. For being in balance does not mean being still, and it does not mean being equal at every place within the system. Dynamic systems are comfortably active within an envelope of balance and it is only when we are close to the edge that we notice big things start to happen.

A QUESTION OF BALANCE

The 15-foot balance pole held steady, hardly affected by the gentle breeze. The bright sunshine carved out sharp shadows, below. As he looked down from the wire 40 feet above the crowd, he realized that a slight movement in either direction would result in disaster. There was a blur of color below him and the oohs and ahs from a small crowd bubbled up in a sign of breathless anticipation. Fighting back the occasional feeling of vertigo and other distractions, his focus was on the wire and his own internal sense of balance. His muscles tensed and relaxed many times each second as he moved his body and the pole ever so slightly to compensate for the wind and the vibration of the wire that sought to send him sailing into the crowd below. He had worked hard to get to this exact spot, and now he strove to maintain it with a continual series of small adjustments.

The view from the ground was quite different. He seemed to move with ease through each routine, and often appeared to stand perfectly still for long periods of time as he acknowledged the applause of the crowd. The thousands of minor movements that allowed him to maintain his position were unseen from just a short distance away.

~ ~ ~

In nature, balance is not a point where everything is still, but rather a stable zone of interdependent relationships within which there is a great deal of activity. Over the long term, the Earth favors balance; it leads to sustainability. Change, on the other hand, can cause a lot of destruction, and if too many species become extinct, life has to start over. There is only so much change we can accept in our lives at any one time.

Movement needs to occur within a sphere of balance. When we are in balance with our neighbors and our surroundings we feel a sense of harmony. A lack of harmony creates stress and weakens the immune systems of all living things and that makes us more vulnerable. In a very real sense, stress is another burden on our systems as destructive as any drug or pollutant. If we spend every hour of every day fighting just to survive, we have no time to think about what we want to be, what we **can** be. In nature, there is a very real advantage for individuals and systems that remain largely in balance.

Earth relationships that move in and out of balance are common and can be looked at on different levels. We find them in the physical processes of the Earth itself and in the relationships between the Earth and the life that calls it home. And there is also a spiritual and emotional balance that we all seek. Let's look at these different relationships and see what happens when things get out of balance.

BALANCE IN THE PHYSICAL PROCESSES OF EARTH

If we look again at the Earth from a great distance, we would surmise there are forces moving the physical Earth out of balance and also processes that act to restore the balance.

The major driving forces leading to imbalance derive from the internal heat left over from the Earth's formation that causes the plates to move and mountains to be built, and additional heat from both natural radioactive decay and the sun. Moving things back to balance are: erosion that chips away at the high places and fills in the low, and the ocean and the atmospheric currents that mix and circulate sending heat where there is cold and rain where it is dry.

Although we have focused our attention on the Earth, it would all seem to start with the sun, a giant 24-hour beacon dissipating raw energy out into space in all directions. It is not selective or judgmental; whatever gets in the way gets a dose. A small portion of that enormous amount of radiated energy and heat falls on the Earth. What isn't retained supporting life by photosynthesis or heating the oceans, air and land is released to space. But much greater heat is collected at the equator—where the angle to the sun is more direct—and without some sort of distribution system, the temperature at the equator would eventually become unbearably hot while the oceans at the poles would freeze solid. Solar heat is distributed away from the equator and toward the poles by strong, established ocean and air currents.

Over the last few years many people have become aware that the Gulf Stream—which originates in the Gulf of Mexico and travels northward past Nova Scotia and on to Europe—brings warmth without which Europe would be vastly colder. If there is any change that causes instability in the Gulf Stream—or worse—causes it to cease, Europe goes into the freezer. There are also ocean currents moving south from the equator that distribute heat to the southern hemisphere and cold currents moving back from both poles to the equator to be heated again. Complex air currents and storms in their own fashion act to bring heated air from the equators, toward the poles. In these ways and with these kinds of currents the world over, a balance is maintained.

In looking at nature's momentous and often catastrophic events, our minds are drawn to images of a planet and its systems gone completely out of balance. But let's start at the other end and examine what it looks like when the Earth restores the balance as it ultimately does. In fact, it is often the things that result from the flow *into* balance that take our breath away and scare us, too. Most of the world's great natural disasters and some of its most magnificent displays of power are not examples of moving further out of balance, but rather of what happens in restoring the balance. There is often a slow and persistent movement away from balance, followed by a rapid and catastrophic event that actually restores the balance. In essence, we see the effect but the cause is more obscure.

NATURE'S TOOLS TO RESTORE THE BALANCE

__Tornados__ - We don't intuitively think of a tornado as an instrument of balance, but that is one role it plays. Occasionally in the spring and summer, in the US, cool, dry air masses flow down from the north above moist rising air that has been warmed from the sun's heat. At this point the Earth is presented with a dilemma: the cool air is heavier and wants to sink; the warm air is lighter and wants to rise, forming an imbalance. The rising turbulent air becomes thunderstorms and will not be denied. At some point there is a break in the boundary between the air layers. The warm air from below rushes up in a rotating motion as the cooler air above pours down through the same area. The movement to restore the balance between the cold and warm air masses has begun and a tornado is born.

When we stop and think about this for a moment we realize a couple of important things. First, all the energy and power of the imbalance that was spread over a hundred square miles of air masses are focused near the spot where balance is restored, the tornado. Second, the building-up of imbalance was relatively slow, forming over the course of at least several hours. But the process to restore the balance was swift and violent.

Hurricanes - So much of what we see on Earth is a game of heat distribution; and hurricanes fall exactly into that category. The Earth receives tremendous heat from the sun each day, less at the poles and much more at the equator. One way the heat is distributed to cooler areas is through cyclonic tropical storms that we call hurricanes in the Atlantic Ocean and typhoons in the northwestern Pacific Ocean.

By mid to late summer, oceans near the equator have absorbed large quantities of heat near the surface that provide the fuel for storm development. Commonly, Atlantic hurricanes begin as easterly waves in the atmosphere that form over West Africa. The waves are disturbances that travel from east to west along the tropical winds. As the waves move out over the ocean they begin to form thunderstorms.

Inside the thunderstorms a cycle develops where warm moist air at the surface rises and cools, causing rain. As the air continues to rise and spill out the top of the storm, adjacent air at the ocean's surface flows in to take its place. If the storm is far enough from the equator, the Coriolis force causes the converging winds to begin to rotate in a counterclockwise motion around the area of lowest pressure and if the winds continue to increase, a hurricane is formed. As it moves north from the tropics it carries tremendous amounts of moisture picked up from the surface of the warm ocean. This moisture, along with large quantities of heat, is eventually released in the form of rain.

The Earth's imbalance in heat near the equator results in giant cyclonic storms that distribute the heat and restore the balance. As with tornadoes, the restoration of balance is powerful, relatively rapid, and violent.

Earthquakes – The surface of the Earth is broken into great crustal plates making up the continents and the oceans. Riding on a molten mantle, there is a movement in all of these plates, but they aren't all going at the same speed. And since there is only one Earth to move around upon, eventually, there has to be a collision. At the boundary of these plates an imbalance is created; it is as if the front of the train has slowed down, but the rear keeps going at the same speed and something has to give. The collision forces one plate to slide beneath the other in a rough and sporadic movement as the pressure continues to build. Eventually an earthquake occurs that relieves the pressure and restores the balance in one rapid and jerking motion. There are some similarities in the earthquake to the tornado: the imbalance builds slowly from pressure over a wide region, but the release is rapid and local, and often destructive.

Lightning - This is a wonderful example of nature's way of restoring electrical balance. The formation of lightning has been studied for hundreds of years, but the exact process that causes a build-up of a negative charge in the clouds above a positively charged Earth is still not completely understood. When the thunderstorm is well-developed and the electrical imbalance between the cloud and the ground becomes too great, we see a lightning bolt of flowing electrons that connects the cloud to the Earth and restores the balance. Lightning is a beautiful thing, unless you happen to be directly in the line of fire.

Natural Fires - Now here is an interesting process to look at. Over a period of time leaves, dead branches, and other debris collect on the floor of our forests. Eventually they will break down in nature's compost and enter the soil. Occasionally a natural fire sweeps through the forest, cleaning out the dead material. Usually this fire moves swiftly along the forest floor and does not involve the trees themselves and the resulting ash enriches the soil for the next period of growth. However in some cases where the build-up of dead material is extreme, the resulting fire can consume large portions of the forest and many of the animals that call it home.

These examples all show that the Earth's physical processes are continually moving in and out of balance much like aspects of our own lives.

BALANCE BETWEEN THE EARTH AND THE LIFE THAT DEPENDS UPON IT

Plants and animals have grown to fit snugly in the environments of Earth and in each case of significant change in temperature, sunlight, moisture, or food source, life has changed in response. As a result of global warming, temperatures in some areas have begun to rise. In addition, weather patterns have shifted causing more floods in some areas, droughts in others. The once dependable climate has changed. Some of Earth's creatures are able to adapt their needs to meet the changing conditions while other life forms of the land and sea have begun to respond by moving away from areas that are no longer suitable to areas that are.

An outbreak of the Hantavirus showed that an imbalance in one area can cause a rippling effect ("Hanta Virus in New Mexico", Dr. Erik McLaughlin, June 13, 2009, General Medicine Community, www.Wellsphere.com). In 1993, in New Mexico, Navajo people fell ill to a disease of frightening proportions with symptoms including high fever, cough, and death. Since it seemed to affect only Navajos, it was first called the "Navajo Flu." But when a visitor from Iceland contracted the disease everyone knew it was not just a disease of the Navajos. In a short time, 11 people had died, including a couple where one partner died in a hospital overlooking the funeral parlor where at the same moment the other was being buried.

An autopsy was eventually performed but the disease was still not identified. The Center for Disease Control (CDC) was called in; but the virus could not be matched against any of the 40,000 samples of known viruses. The CDC also talked with the Navajo elders who said they knew of this disease, that it had appeared in the 1930s, in 1914, and in the last century. It had always appeared after two seasons of unusual rain, and mice were always in abundance. They believed the mice were connected to the disease, but didn't know exactly how.

Based on the data from the elders, the CDC focused on diseases with similar symptoms and associations with mice. They even considered the Bubonic Plague that killed millions in the middle ages and was carried by fleas on rats; but it was not the Bubonic Plague.

Finally a test comparison of antibodies was run. When the body is exposed to a virus, it produces antibodies that correspond to the disease. Although comparisons found no direct match, there were similarities to a virus that caused the death of about 400 soldiers in the Korean War. The virus came from the Hun Tan Valley, and was carried by mice and was named the Hantavirus after the Hantan River. The mice carry and shed the virus in their fecal material and in their urine. As it dries it becomes dust and is airborne and very contagious.

The Navajo associated the disease with a time when there was unusually heavy rains for two seasons, and a particular pine tree produced high quantities of nuts. The deer mice came to feed on the nuts and were multiplied, and the deaths began. Often an imbalance in nature causes a rippling effect through many species. So one imbalance causes another and then another and so on.

Oceans represent another complex system; they are all connected, but they are not all alike. While surface currents formed by the winds are important, it is bottom currents that drive the productivity of life in the world oceans. As ice is formed in the great shelves of the Arctic and Antarctic, it leaves the salt behind so the water below the ice gets even saltier. Salty water is heavier water and it sinks, providing a source of energy for bottom currents around the world. They are not only heavy and cold; they are also very rich in nutrients. When bottom currents collide they often move upward, bringing nutrient-rich waters near the surface where plankton consume the nutrients and multiply rapidly. But it doesn't stop there; other organisms in the food chain feed on the plankton and then others feed on them and so on. Even the great whales come to feed in what is one of the biggest free lunches on the planet. A disturbance anywhere in the chain affects the balance of the whole system and the rippling effect goes on and on. The same is true of all the connected components of our lives including our relationships.

SPIRITUAL AND EMOTIONAL BALANCE

Today our world is deeply painted in the colors of conflict, from the level of nations right on down to the level of the family. The culture reflects the individual. Physical violence often has its first seeds in the harsh tone and the abusive word and it can be a response to slow-building everyday internal stresses and pressures. The high demands and expectations of daily life often produce stress that can take people out of their comfort zone, out of their *zone of balance*. The response is all too often a regrettable eruption that reinforces the pattern of violence and leaves scars on all those around.

What lessons can be learned from the examples of the way balance is restored in nature? The movement out of balance occurred slowly while restoring the balance was fast and powerful, with little warning that we could observe; and the energy was focused in relatively small areas.

Our lives can mirror nature as well. Often stress and anger that has originated slowly over wide areas of our lives become focused in just one place and the result can be violence or abuse. Imagine the person who has a difficult day at work—stress—then gets away late only to be stuck in a grid of traffic—more stress. Perhaps another driver cuts him off with an epitaph—still more stress. By the time the person gets home he is very much out of emotional balance. It takes only the slightest additional irritation and the person may explode and kick a pet, slap a child or hit a mate. So what was built up slowly through the course of the day is released quickly and sometimes violently.

Natural fires remove dead material from the forests and there is a lesson in this process. It is one thing for a forest to receive a spark from lightning, but a rampaging fire only occurs when there is dead underbrush that has accumulated over a period of years. How much "dead underbrush" do we carry in our own lives, the things we should have cleaned out and let go of long ago? The sparks of an emotional ignition are all around us, we experience them every day. Periodically we need to close old debts and let go of resentments and grudges that can cause us to "burst into flame" with the slightest spark of provocation.

Forgiveness is the tool we use to cleanse ourselves and let go of the decaying things that gnaw at us and diminish our happiness. It is one of the hardest yet most rewarding things we can do for ourselves. Forgiveness releases and allows the one who forgives to live freely in the moment. *We need to forgive.*

In looking at the Earth, we said that the effect is obvious, the cause, more obscure. This is true in people as well. The addictions we see around us whether drugs or too much work are symptoms, not the cause. The angry outburst or the tearful sob releases and restores a temporary balance; but unless life changes are made, the stress continues to build and more eruptions will follow. The balance we seek is lasting and is built on harmony and peace.

Maintaining balance in our emotional and physical lives is crucial to developing a life of peace and nonviolence. In our lives we look for tools, little things or big things that will help us implement our decision for nonviolence each day. As we look to those leaders and prophets that have been able to lead and teach a life of nonviolence we see a common characteristic: each was able to maintain a harmonious physical and emotional balance in their own lives. It was from this centered place that they began each day and it was from this foundation that they lived. When each was tested by some difficult, unfair and often painful experience they did not have to reach and struggle for calm and a nonviolent response. For the most part they simply pulled up a bucket from their wellspring of peace and spread it to those around them.

Periodically, the great leaders needed to get away from the crowds to contemplate and to re-center. In most cases they chose a quiet secluded place where they were surrounded by the rhythms of nature's centering balance. Here they reminded themselves of the connectedness of all things and retuned themselves to the natural rhythms of Earth.

We require both physical and emotional nourishment to sustain our lives in harmony and peace. In the United States we often take our food for granted. We look forward to the next meal and are certain of its arrival. Yet in some distant and often unrecognized portion of the brain we know that eating is more than a pleasurable experience. Our body reminds us of our needs with hunger pains that sit in the front of our awareness because on some level it knows that without food we will eventually die. We also require spiritual nourishment, but we are not always aware of the signs of hunger. If we do not recognize the signals they continue until there is a sore spot on the Soul. Recently, we have shown a great concern regarding our diets, but I can tell you that many of us are not eating right spiritually.

We know that stress is one factor that can cause us to feel out of balance. And our own life exploration and growth also causes us to leave our comfort zones and explore, although we might initially find it uncomfortable. The deep-seated need for spiritual evolution will not be swayed and with each new enlightenment, each step of the way, a new balance is achieved.

Is being out of balance always bad? Learning occurs when we are out of balance so the answer is no, being out of balance isn't always bad; in fact it's inevitable. We cannot force our way into balance, when we develop the proper conditions in our lives, we *flow* into balance.

Growth and transformation often occur in moments of epiphany, not in a slow and linear progression. Like steps in a staircase, each step of spiritual growth forms a new zone of balance that allows us to gain our footing, then measure and prepare for the next step to a higher level.

Seven simple values for complicated and evolving lives. The same principles that act to sustain the physical Earth, that caused the formation of life and nurtured its close relationship to Earth, act in our own physical and spiritual lives. Through all the challenges and all the unfathomable changes to come, the slow hand of nature continues to guide and direct to new discoveries of the mind and the heart.

There is wonder and splendor, mysticism, vibration, energy, and above all, a smile in nature as big as the sunset. When we commune with the Earth we are standing in the presence of Creation and we are all invited to sit down and enjoy the view.

EPILOGUE

The days continue to pass, the Earth moves again around the sun as the stars slowly rearrange their ancient patterns. And me, I'll keep poking around my little Dinosaur Ridge when I can and look for new discoveries under a sky that bathes the Ridge in different hues and different wraps—the rain, the snow, and the sunshine. For every day on the Ridge is a good day; it anchors me in spirit, maps my journey and centers me in nature's majesty. I am invigorated and alive to know that I am somewhere in deep space, moving along quite comfortably with the most beautiful planet one could ever imagine. Lead on Planet Earth, you are in my blood and in my genes.

ABOUT THE AUTHOR

ADAM CHAPUIS

There are always certain places and events, certain people that shape a life — some leave imprints in big bold letters, some in more subtle but permanent influences.

Adam was raised on a small farm in Alabama. The early years were laced with fields of red clover and the smell of fertilizer in the summer, and there was hay along with cotton seed hulls and meal for the cows in the winter when there was no grass. His family had a large garden for vegetables which was common in their neighborhood and his father always had cows, in the early years for milk and cream and to sell the steers.

The first house he remembered was small with four rooms and a tin roof. It had a simple fireplace that had been boarded up and a gas heater installed which was a good thing since a snake lived in the old fireplace too for the shelter and the warmth in winter. It was what his parents called a spreading Adder, yellow with black. They used to find it out in the yard occasionally until the dog found it first, then no more snake.

He spent many days and nights in the company of a few friends seeking treasures and adventure like all boys. These friends, along with his sister, Helen formed his universe for the early years.

As boys he and his friends wandered through what they thought were deep woods, thick forests of oak, hickory, maple, sweetgum and countless other trees sprinkled with the occasional pine. They explored on foot and on bicycle and sought to know every inch of the land at every time of day and night. When he was older, they did most of their exploration on horseback, not really because they felt like horsemen, but because you can pretty well go anywhere on a horse. His family's first horse was sleek and young and he thought she was imposingly beautiful. Later he wondered what part of hell she was born in. The problem was that she was very smart for a horse, she understood ritual and routine and it didn't take her long to figure out the entire family both as individuals and as a group. Individually she saw in Adam a scared little kid pushed by his father to ride the great beast; in other words something to easily torment.

Somewhere around the age of thirteen in yet another summer of thick air and insects, lightning bugs and black-eyed peas he noticed that many of their neighbors but a few miles away maintained dairy farms, and the rich black soil of their farms grew alfalfa and other silage for their cows. He remembered looking for that rich black soil on their own land and finding none. He didn't understand then that their farm was on the flood plain of the Alabama River, nested in acidic red clayey soil that only welcomed and raised the likes of squash and green beans: watermelons and cantaloupes need not apply. They were so near yet so far away from that precious soil of the Black Belt that ran from Selma through

Montgomery. He didn't know it then but that soil grew dreams, in the same way it turned the imagination of any seed into full bloomed fruit. This was the first time he could remember wondering about the Earth and he began to notice things, the shape of the land along the creek banks, and where the wild plumbs grew. It had not yet struck him that the precious soil of Earth was a medium that grew whatever was placed in it without question, without judgment.

He enjoyed the sciences in high school, not the studying of them mind you, but the idea of them and the physical experiments. He loved chemistry because they built rocket cars and in moments of boredom they chased and squirted each other with bottles of hydrochloric acid. He often wondered if his mother ever questioned where the holes in his shirts came from; she never said anything. When he arrived on campus for his undergraduate days, everything changed. Chemistry became inordinately "un-fun" and a series of life changes moved him to the study of geology which he recognized almost immediately as a true calling. After a Masters degree, he spent a number of years in various fields eventually directing a small Environmental Engineering and IT services firm.

The author is a hiker, camper, bicycler, snorkeler, and diver, and enjoys playing guitar and African Drums. These days he shares his time between Colorado and Key West, enjoying those places where the beauty and power of Earth are easily touched. Being an Earth Monk is a state of mind and it is the way he perceives himself in his truest form.

TO LEARN MORE ABOUT
VOICES OF THE EARTH
&
ADAM CHAPUIS

Adam Chapuis is a speaker and workshop leader. He is available for speaking engagements and personal appearances and may be contacted at **rachapuis@EarthMonk.com**

RECOMMENDED READING

A New Earth: Awakening to Your Life's Purpose
Tolle, Eckhart   ISBN: 0452289963

How to Know God: The Soul's Journey into the Mystery of Mysteries
Chopra, Deepak, M.D.   ISBN: 0307407748

Touch the Earth: A Self Portrait of Indian Existence
McLuhan, T.C.   ISBN: 0883940000

Daedalus Rising: The True Story of Icarus
Case, Robert William   ISBN: 0982083815

The Power Of Myth
Joseph Campbell, Bill D. Moyers -   ISBN: 0385418868

The Fabric of the Cosmos: Space, Time and the Texture of Reality
Greene, Brian   ISBN: 0375727205

EarthDance: Living Systems in Evolution
Sahtouris, Elisabet   ISBN: 0595130674

The Artist's Way: A Spiritual Path to Higher Creativity
Cameron, Julia   ISBN: 1585421472

Mindful Drumming: Ancient Wisdom for Unleashing the Human Spirit and Building Community
Clottey, Kokomon   ISBN: 0971967806

If you liked
Voices of the
Earth
Please leave a
review on
Amazon.com

Also available in
Kindle